Process Development Division

AA003363

Core Programming Topic at the 2011 AIChE Spring Meeting and 7th Global Congress on Process Safety

Chicago, Illinois, USA
13-17 March 2011

ISBN: 978-1-61782-904-8

Printed from e-media with permission by:

Curran Associates, Inc.
57 Morehouse Lane
Red Hook, NY 12571

CURRAN ASSOCIATES INC. proceedings.com

Some format issues inherent in the e-media version may also appear in this print version.

Printed by Curran Associates, Inc. (2011)

For permission requests, please contact AIChE
at the address below.

AIChE
3 Park Avenue
New York, NY 10016-5991

Phone: (203) 702-7660
Fax: (203) 775-5177

www.aiche.org

Additional copies of this publication are available from:

Curran Associates, Inc.
57 Morehouse Lane
Red Hook, NY 12571 USA
Phone: 845-758-0400
Fax: 845-758-2634
Email: curran@proceedings.com
Web: www.proceedings.com

TABLE OF CONTENTS

This welcoming session sponsored by Fuels & Petrochemicals Division (F&PD) is a new initiative organized by the AIChE's national and F&PD technical programming team as a preamble to the division's sponsored technical sessions. This will be opened by a short introduction by the chair of the division. Besides recognizing leadership in his introductory short presentation, the chair will highlight the key elements of value additions the division brings to its members. The presentation will be concluded with a special request to join the division where professional opportunities are limited only by members' imagination, to positively impact academia and industry and thereby society at large. For details of the division and its many facets of activities and functions please follow the link www.aiche-fpd.org.

(10b) What Are We Going to Do with All This Natural Gas? - Ramifications of the Dramatic Expansion of Natural Gas Available to the Market Due to North American Shale Gas Production and LNG Liquefaction

Sam Thomas, Chart Industries, Inc., Garfield Heights, OH

The success in Shale and other unconventional formations has dramatically expanded the N. American Natural Gas reserves and promises to also do so in many other parts of the world. New LNG liquefaction capacity in the Middle East and Asia along with additional large scale capacity from Australia and Papua New Guinea coming on-line in the next 5 years opens the opportunity for a global trade in Natural Gas stabilizing historically volatile pricing. How will this important energy and petrochemical feedstock be utilized? The success in improving techniques to allow unconventional gas reserves to be proven and significantly increase the available reserves has not as yet been fully appreciated by markets and policy makers. This presentation reviews the ramifications of natural gas becoming a larger part of the global energy supply for which there are numerous competing interests. Natural Gas holds the ability to significantly reduce carbon emissions by substitution for coal in power generation, it can be used as a transportation fuel either as LNG, CNG or GTL clean diesel increasing energy independence of some of the largest economies in the world simultaneously reducing carbon emissions. It could enable the USA to maintain its position as a leading petrochemical producer if used for ethylene feedstock. How will these competing interests play out and how will governments respond?

(14a) The Energy Challenge: Providing Sustainable Supplies to Meet Growing Demand

F. Emil Jacobs, ExxonMobil Research and Engineering Company, Annandale, NJ

As economies grow and living standards improve, the world faces a dual energy challenge: meeting the growing demand for energy while mitigating emissions and their impact on the environment. ExxonMobil believes that addressing this dual challenge requires an integrated set of solutions which includes increasing the supply of all commercially viable energy sources, reducing emissions, and improving energy efficiency in all sectors of the economy. Technology will be a key enabler of continued progress in all of these areas.

Many experts agree that the immense scale of global energy demand necessitates a continuing role for oil and gas as a substantial portion of the energy supply mix for at least the next several decades. These experts also forecast that the overall energy mix will shift toward a relatively higher level of sustainability through a combination of increased efficiency across the economy, growth in lower-carbon sources of energy, and development and deployment of technologies that help mitigate emissions.

ExxonMobil has developed a number of technologies that contribute to increased supplies of energy, reduced emissions, and higher efficiency in its operations and in consumer use of its products. The presentation will provide an overview of ongoing research and development activity in a broad range of technology areas including algae-based biofuels. It will also highlight some of company's current technologies including lithium-ion battery separator films, lightweight plastics that improve automotive efficiency, and processes that could advance carbon capture and storage.

(31a) Upgrading Bed Support Systems Utilizing Vee-Wire Technology

Benjamin Schmitt P.E., Hydrocarbon Processing Industry Business Unit, Johnson Screens, New Brighton, MN

Abstract:

Upgrading Bed Support Systems, Utilizing Vee-Wire Technology

By: Benjamin Schmitt, PE

Date: October 18, 2010

Hydrocarbon industry processes utilize media (sieve or catalyst) to catalyze a process or adsorb impurities and other contaminates. Support structures are an integral part of the process used to prevent the media from passing down-stream creating problems with mechanical devises or impact end product specifications. Media retention structures vary from mesh over grating or perforated plate to Vee-Wire screen. A technical review of the advantages for utilizing Vee-Wire technology will be presented, sighting case studies and supporting data. Topics of discussion will include process improvements, mechanical advantages, service life and capital recovery of Vee-Wire support structures.

Theoretical Modeling of a Novel Combination of Pd-Ag Membrane Reactors in Methanol Synthesis

A. A. Forghani, M. Bayat, M. R. Rahimpour, School of Chemical and Petroleum Engineering, Shiraz University, Shiraz, Iran

Introduction

Fundamentally, methanol is produced by catalytic conversion of synthesis gas (CO_2, CO and H_2). The factors affecting the production rate in industrial methanol synthesis are parameters such as thermodynamic equilibrium limitations and variation in stoichiometric number. For equilibrium reactions, selective product removal or reactant addition may be used to increase conversion. Up to the maximum production rate point, increasing temperature improves the rate of reaction, which leads to more methanol production. Nevertheless, as the temperature increases beyond this point, the deteriorating effect of equilibrium conversion emerges and decreases methanol production [1]. Rahimpour proposed a two-stage catalyst bed concept for conversion of carbon dioxide into methanol [2]. Rahimpour et al. have investigated enhancement of methanol production in membrane dual-type reactor [3] and recently Rahimpour et. al. investigate a novel single membrane reactor in methanol process for increasing methanol production [4]. The main goal of this work is the improvement of methanol production in dual-type methanol reactors. In this new system, the walls of tubes in the both of reactors are coated with a hydrogen perm-selective membrane. The application of membrane reaction technology in chemical reaction processes are now mainly focused on reaction systems containing hydrogen and oxygen, and are based on inorganic membranes such as Pd and ceramic membranes [5]. The hydrogen partial pressure gradient is the driving force for hydrogen permeation from feed synthesis gas to the reacting gas. The advantages of this concept will be discussed based on temperature and concentration profiles along the reactors. On the other hand proficiency of conventional dual-type methanol reactor (CDMR) is used as a basis for comparison in terms of methanol production, temperature of gas phase. This comparison shows that the production rate of cascade membrane methanol reactor (CMMR) is greater than CDMR. Moreover due to favorable temperature profile along this system is suitable for catalysts life time that used in aforementioned synthesis.

Process Description

Conventional Dual-type Methanol reactor (CDMR)

Overall shape of this reactor was similar to MDMR in Fig. 1. The catalyst is packed in vertical tubes of first reactor and shell side of second reactor. The technical design data of the catalyst pellet and input data of CDMR are summarized in Tables 1 and 2. This system is mainly based on the two-stage reactor system consisting of a water-cooled and a gas-cooled reactor. The cold feed synthesis gas is fed to the tubes of the gas-cooled reactor (second reactor) from bottom of reactor and flowing in counter-current mode with reacting gas mixture in the shell of the reactor. Then the synthesis gas is heated by the heat of reaction produced in the shell. Therefore, the reacting gas temperature is continuously reduced through the reaction path in the second reactor. The outlet synthesis gas from the second reactor is fed to tubes of the first reactor (water-cooled) and the chemical reaction is initiated by the catalyst. The heat of reaction is transferred to the cooling water inside the shell of reactor. In the first stage, methanol is partly produced. The gas leaving the first reactor is directed into the shell of the second reactor from top of the reactor. Finally, the product is removed from the downstream of the second reactor (gas-cooled reactor).

Table 1. Catalyst and specifications of both reactors of CDMR [6]

Water-cooled reactor		Gas-cooled reactor	
Parameter	Value	Value	Unit
D	4.5	5.5	[m]
D_i	40.3	21.2	[mm]
D_o	4.5	25.4	[mm]
d_p	0. 00574	0.00574	[m]
ρ_s	1770	1770	[kg m^{-3}]
Tube length	8	10	[m]
Number of tubes	5955	3026	[-]
Shell side pressure	-	71.2	[bar]
Tube side pressure	75	76.98	[bar]

Table 2. Input data of the CDMR [6]

Feed conditions	Value
Feed composition (mole %):	
CO_2	8.49
CO	8.68
H_2	64.61
CH_4	9.47
N_2	8.2
H_2O	0.1
CH_3OH	0.37
Inlet Temperature[K]	401

Cascade Membrane Methanol Reactor (CMMR)

Fig. 1 shows the schematic diagram of a cascade membrane reactor configuration for methanol synthesis. The methanol synthesis process of cascade membrane methanol reactor is similar to that in the dual-type methanol reactor, with the exception that in tubes walls in the first reactor and second reactor consist of hydrogen-permselective membrane and pressure difference between the shell and tube permits diffusion of hydrogen through the Pd-based membrane layer. Therefore, the mass and heat transfer occurs simultaneously between shell and tube. This simulation study is based on a Pd–Ag layer thickness of 1.1 μm. In this configuration, the outlet synthesis gas from gas cooled reactor is fed to the tube side (permeation side). Pure hydrogen permeates to reaction side in order to control and maintain the suitable hydrogen gradient in the whole length of the reactor. Then, the heated synthesis gas is routed from recycle stream through shell in a co-current mode with feed synthesis gas. The reacting gas is also cooled with cooling saturated water which flows around it. In fact, the heat of reaction is transferred to both cooling water and outlet synthesis gas from gas cooled reactor. The gas leaving from shell side of first reactor is directed into the shell of the second reactor from top of the reactor. Finally, the product is removed from the downstream of the second reactor (gas-cooled reactor). The input data and some of the technical design data are the same as CDMR (see Table 2).

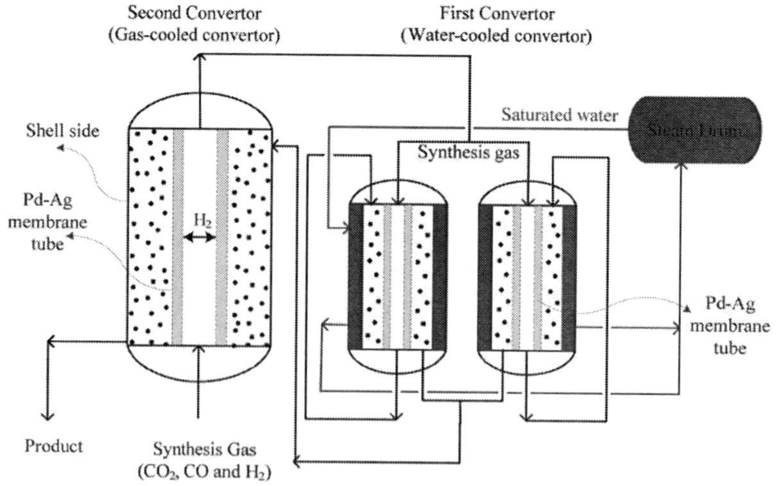

Figure 1. Schematic diagram of a cascade membrane methanol reactor (CMMR)

Mathematical model

Water-cooled reactor (first reactor)

The first reactor has been modeled regarding the following assumptions:
(a) One-dimensional plug flow; (b) Axial dispersion of heat is negligible compared to convention; (c) Gases are ideal. (d) The radial diffusion in catalyst pellet is neglected. (e) There are no radial concentration and temperature gradients. (f) Steady state condition. The following equations are written for fluid phase:
The mass and energy balance equations for the catalyst pellets can be formulated as follows:

$$k_{g_i} a_v . c_t .(y_i - y_{is}) + \eta r_i \rho_B a = 0 \qquad i = 1,2,..., N-1 \qquad (1)$$

$$a_v h_f (T - T_s) + \rho_B a \sum_{i=1}^{N} \eta r_i (-\Delta H_{f,i}) = 0 \qquad (2)$$

where, y_{is} and T_s are the mole fractions on the catalyst surface and solid-phase temperature, respectively. As can be seen in Fig. 1, the outlet synthesis gas from the second reactor is the inlet synthesis gas to the first reactor. The boundary conditions are unknown and the more details are explained as numerical solution in steady state computations.

$$z = 0; \quad F = F_{in}, \; y_i = y_{i,in}, T = T_{in} \qquad (3)$$

The mass and energy equations for the bulk gas phase can be written as follows:

$$-\frac{1}{A_c}\frac{\partial F_i}{\partial t} + a_v c_t k_{gi}(y_{is} - y_i) + \frac{\alpha_H}{A_s}(\sqrt{P_H^{\,t}} - \sqrt{P_H^{\,sh}}) = 0 \qquad i = 1,2,..., N-1 \qquad (4)$$

$$-\frac{1}{A_{shell}} C_{pg} \frac{\partial (F^{sh} T)}{\partial z} + a_v h_f (T_s - T) + \frac{\alpha_H}{A_s}(\sqrt{P_H^{\,t}} - \sqrt{P_H^{\,sh}}) c_{ph} (T_{tube} - T) + \frac{\pi D_i}{A_{shell}} U_{tube} (T_{tube} - T) + \frac{\pi D_o}{A_{shell}} U_{shell} (T_{shell} - T) = 0 \qquad (5)$$

Tube side (Permeation side):

The mass and energy balance equation for fluid phase are given:

$$-\frac{1}{A_c}\frac{\partial F_i^{\,t}}{\partial z} - \frac{\alpha_H}{A_c}(\sqrt{P_H^{\,t}} - \sqrt{P_H^{\,sh}}) = 0 \qquad i = 1,2,..., N-1 \qquad (6)$$

$$-\frac{1}{A_c} C_{pg} \frac{\partial (F^t T)}{\partial z} + \frac{\alpha_H}{A_c}(\sqrt{P_H^{\,t}} - \sqrt{P_H^{\,sh}}) C_{ph} (T - T_{tube}) + \frac{\pi D_i}{A_c} U_{tube} (T - T_{tube}) = 0 \qquad (7)$$

where, F^t is molar flow rate. When α_H is zero, the membrane is not permeable to hydrogen and the model is used for conventional dual-type methanol reactor (CDMR).

The boundary conditions are as follow:

$$z = L; \quad y_i = y_{if}, \quad T = T_f \qquad (8)$$

Gas- cooled reactor (second reactor)

Shell side (reaction side)

The mass and energy balance for solid phase in the gas-cooled reactor is the same as that in the water-cooled reactor. The following equations are written for fluid phase:

$$-\frac{1}{A_c}\frac{\partial F_i}{\partial t}+a_v c_t k_{gi}(y_{is}-y_i)+\frac{\alpha_H}{A_s}(\sqrt{P_H{}^{t}}-\sqrt{P_H{}^{sh}})=0 \qquad i=1,2,\ldots,N-1 \qquad (9)$$

$$-\frac{1}{A_{shell}}C_{pg}\frac{\partial(F^{sh}T)}{\partial z}+a_v h_f(T_s-T)+\frac{\alpha_H}{A_s}(\sqrt{P_H{}^{t}}-\sqrt{P_H{}^{sh}})c_{ph}(T_{tube}-T)+\frac{\pi D_i}{A_{shell}}U_{tube}(T_{tube}-T)=0 \qquad (10)$$

where, y_i and T are the mole fraction of component i and gas-phase temperature, respectively. F^{sh} is molar flow rate, c_{pH} is specific heat of the gas at constant pressure, h_f gas-catalyst heat transfer coefficient and α_H Hydrogen permeation rate constant and $P_H{}^{t}$ and $P_H{}^{sh}$ Tube side pressure and Shell side pressure respectively.

Tube side (Feed synthesis gas flow)

The mass and energy balance equation for fluid phase are given:

$$-\frac{1}{A_c}\frac{\partial F_i^{t}}{\partial z}-\frac{\alpha_H}{A_c}(\sqrt{P_H{}^{t}}-\sqrt{P_H{}^{sh}})=0 \qquad i=1,2,\ldots,N-1 \qquad (11)$$

$$-\frac{1}{A_c}C_{pg}\frac{\partial(F^{t}T)}{\partial z}+\frac{\alpha_H}{A_c}(\sqrt{P_H{}^{t}}-\sqrt{P_H{}^{sh}})C_{ph}(T-T_{tube})+\frac{\pi D_i}{A_c}U_{tube}(T-T_{tube})=0 \qquad (12)$$

where, F^t is molar flow rate. When α_H is zero, the membrane is not permeable to hydrogen and the model is used for conventional dual-type methanol reactor (CDMR).
The boundary conditions are as follow:

$$z=L; \quad y_i=y_{if}, \quad T=T_f \qquad (13)$$

Results and discussion

Figure 2 (a) compares methanol production rate profile along the two types of reactors. Due to overcome mass and heat transfer limitations as well as shifting the equilibrium by using perm-selective membrane in CMMR system, this reactor configuration proves a considerable enhancement in methanol production rate in comparison with CDMR system.

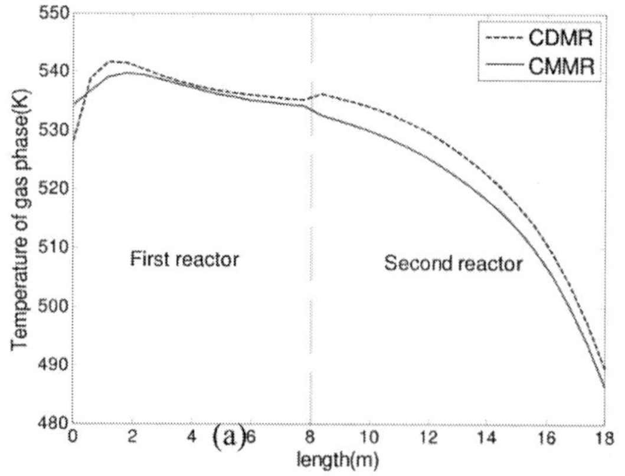

(a)

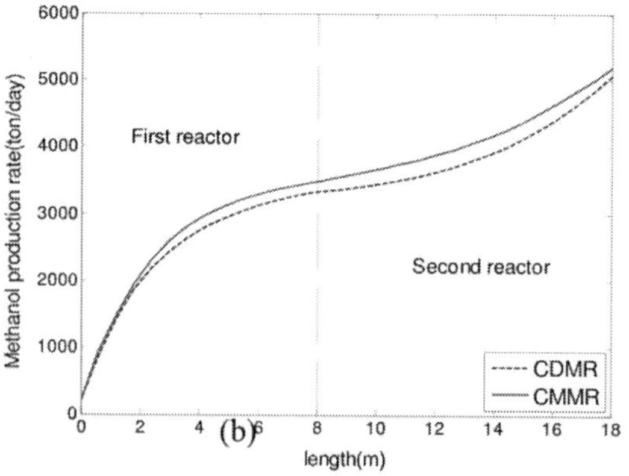

(b)

Figure 2. Comparison of methanol production rate profile (a) and Temperature of gas phase (b) in CMMR and CDMR systems

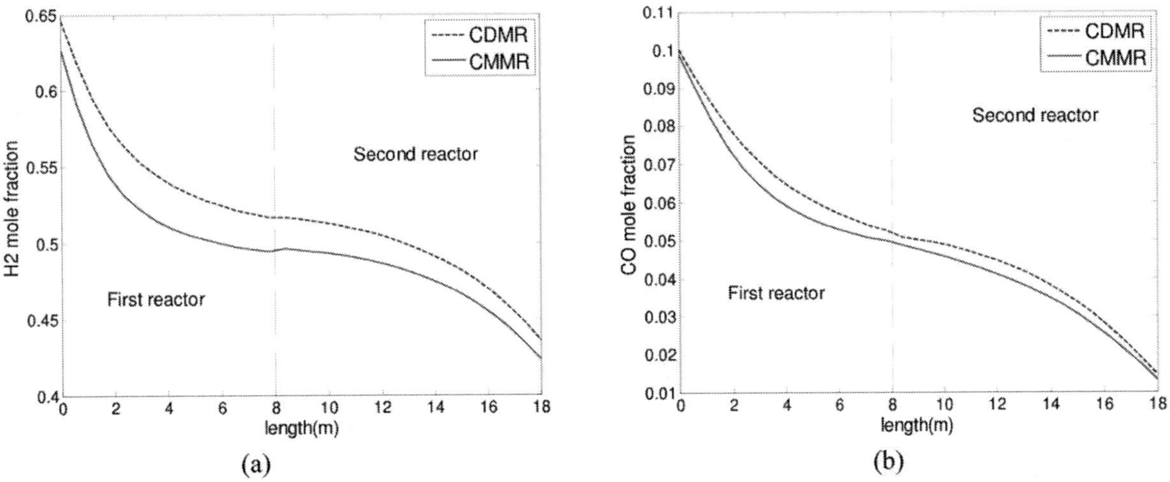

Figure 3. Comparison of CO mole fraction (a) and H2 mole fraction (b) in CMMR and CDMR systems

Figure 2 (b) demonstrates gas phase temperature in two CDMR and CMMR systems. According to this figure more favorable temperature along CMMR configuration achieved and this system is suitable in methanol production because variation of temperature is less than CDMR system and catalyst deactivation achieved in longer time than CDMR system.

Figure 3 shows reactants mole fractions along the two types of CDMR and CMMR configurations. According to this figure more reactants convert to methanol in CMMR system in comparison to CDMR system, so more methanol produced in CMMR because of H_2 diffusion through membrane in first and second reactor.

Conclusion

The methanol forming reactions are strongly exothermic and limited by the thermodynamic equilibrium. Therefore, the development of a membrane-based process could open the way to increasing the methanol production in the methanol synthesis process. In this work, cascade membrane reactor system was compared with a conventional dual-type reactor for methanol synthesis. The potential possibilities of the cascade membrane reactor system were analyzed using a one-dimensional heterogeneous model to obtain the necessary comparative estimates. A comparison of calculated temperature profile along the reactor length for two systems shows the extremely favorable temperature profile over the catalyst beds of cascade membrane methanol reactor and it is useful for methanol catalysts from deactivation feature. This feature suggests that the concept of cascade membrane dual-type reactor system is an interesting candidate for application in synthesis gas conversion to methanol.

References

1. Rahimpour M. R., Ghader S., Baniadam M., Fathikalajahi J., Incorporation of flexibility in the design of a methanol synthesis loop in the presence of catalyst deactivation. *Chem Eng Technol* 2003; **26 (6)**:672-678.

2. Rahimpour M. R., A two-stage catalyst bed concept for conversion of carbon dioxide into methanol. *Fuel Process Technol* 2008; **89(5):**556-566.
3. Rahimpour M. R., Lotfinejad M., Enhancement of methanol production in a membrane dual type reactor. *Chem Eng Technol* 2007; **30**:1062-1076.
4. Rahimpour M. R., Khosravanipour Mostafazadeh A., Barmaki M. M., Application of Hydrogen-Permselective Pd-based Membrane in an Industrial Single-type Methanol Reactor in the Presence of Catalyst Deactivation. *Fuel Process Technol* 2008; **89(12)**:1396-1408.
5. Rahimpour M. R., Asgari A., Modeling and simulation of ammonia removal from purge gases of ammonia plants using a catalytic Pd–Ag membrane reactor. *J Hazard Mater* 2008; **153**:557-565.
6. Domestic industrial methanol reactor, Design manual; 2007.

(31c) Benzene Hydrogenation - Achieving Process Heat Integration and Intensification

William Cross Jr., CDTECH / Chemical Research & Licensing, Pasadena, TX and Arvids Judzis, CDTECH / Chemical Research & Licensing

Choices between plant capital and plant energy efficiency are ever present in our industry. The hydrogenation of benzene offers a good example of these trade-offs, which are made by process design engineers daily. The current application of benzene hydrogenation process technologies and how they recover heat will be discussed. Simple flowsheet modifications for existing technology will be analyzed, and more heat integrated process options will be compared. Pros and cons to the various choices will be discussed and conclusions drawn for the two primary industrial applications of benzene hydrogenation.

(31d) New Application of Ionic Liquids as Inhibitors for Flow Assurance

Chongwei Xiao, Petroleum Recovery Research Center (PRRC), New Mexico Institute of Mining and Technology, New Mexico, NM and Hertanto Adidharma, Department of Chemical & Petroelum Engineering, University of Wyoming, Laramie, WY

Hydrates are crystalline solids composed of hydrogen-bonded water molecules and gas molecules. Hydrates tend to form in gas pipelines and drilling fluids when there are sufficient reactants in situ, together with the necessary pressure and temperature conditions. The fluid flow blockage by hydrate in pipelines is a serious problem in natural gas transportation. To assure fluid flow, thermodynamic and kinetic inhibitors are widely used by industry to prevent hydrate formation and thus guarantee gas transportation. However, the existing inhibitors are highly uneconomical, especially at high pressures and high degrees of supercooling.

A preliminary study of imidazolium ionic liquids as a new class of inhibitors demonstrated that certain types of ionic liquids are found to shift the hydrate dissociation/stability curve to a lower temperature and slow down the hydrate nucleation rate, which is attributed to their strong electrostatic charges and hydrogen bond with water. This dual function is expected to make these novel inhibitors perform more effectively than the existing inhibitors. This current research extends the field of ionic liquids to dialkylimidazolium-based ionic liquids with halide anions. Dialkylimidazolium ionic liquids with several alkyl substitute groups in the cation heads and anions have been investigated for their potential application as novel gas hydrate inhibitors. Their effects on the equilibrium methane hydrate dissociation curve in a pressure range of 36 to 136 bar and the induction time of methane hydrate formation at 114 bar and a high degree of supercooling, i.e., about 25°C, are measured in a high-pressure, micro-differential scanning calorimeter. In order to clarify the mechanism of these ionic liquids in inhibiting the hydrate formation, the electrical conductivity and infrared spectra of ionic liquids were measured. The strength of hydrogen bonding between water molecules and anions/cations of the ionic liquids were analyzed from infrared spectra absorption.

In the pressure range of our experiments, the ionic liquids studied with concentrations of 10 wt% shift the equilibrium methane hydrate dissociation curve about 1-3.5 K to lower temperatures. Among all the ionic liquids studied, EMIM-Cl is the most effective thermodynamic inhibitor. For ionic liquids with the same cation, either EMIM or BMIM, the thermodynamic inhibition effectiveness is decreasing along the same sequence: Cl > Br > I > BF_4. For ionic liquids with the same anion, the thermodynamic inhibition effectiveness of ionic liquids with a shorter alkyl chain substituent is better than that of ionic liquids with a longer alkyl chain substituent. The effect of the anion type on the thermodynamic inhibition effectiveness is arguably more significant than that of the cation type. The electrical conductivity measurements demonstrate that the thermodynamic inhibition effectiveness of ionic liquids is closely related to the electrical conductivity of ionic liquids in aqueous solutions. Ionic liquids with higher electrical conductivity for their aqueous solutions generally show higher thermodynamic inhibition effects. The IR spectra of ionic liquids also suggest that the strength of the hydrogen bond with water may have an important role in the thermodynamic inhibition effectiveness of the ionic liquids. From the study of the induction time of hydrate formation, the kinetic inhibition effect of BMIM-I is as good as EMIM-BF_4. The mean value of the induction time of methane hydrate formation from samples containing 1 wt% BMIM-I is about 5.2 hours, compared to 5.7 hours for 1 wt% EMIM-BF_4. Thus, BMIM-I is also found to be much better than commercial kinetic inhibitors such as Luvicap and PVCap. In summary, these new ionic liquids are found to shift the equilibrium hydrate dissociation/stability curve to a lower temperature and simultaneously retard the hydrate formation by slowing down the hydrate nucleation rate.

A Theory of Acoustics in Solar Energy

Himanshu Dehra
Noise Behaviour Institute, McGill University, Montréal, Québec, Concordia
Department of Public Health Medicine, McGill University, Montréal, Québec, Concordia
Energy Management Research Centre, McGill University, Montréal, Québec, Concordia
Department of Mining Engineering, McGill University, Montréal, Québec, Concordia
Department of Building Engineering, Concordia University, Montréal, Québec, Concordia
Centre for Building Studies, Concordia University, Montréal, Québec, Concordia
Concordia Institute of Information Systems Engineering, Concordia University, Montréal, Québec, Concordia
Départment de Communication Sociale et Publique, UQÀM, Québec, Concordia
Akal Takht (1-140 Avenue Windsor, Lachine, Québec H8R 1P7) Concordia
* Corresponding author. Tel: +1 514 3663982, Fax: +1 514 3721918, E-mail:anshu_dehra@hotmail.com

Abstract
A theory of acoustics in solar energy has been projected. An amplifier constituting of a parallel plate photovoltaic device connected to a potentiometer has been built. It was constructed with a pair of glass coated photovoltaic modules and polystyrene filled plywood board as back panel with air ventilation through a parallel plate channel of amplifier. The sample results obtained from experiments and simulation model are presented to support hypothesis of acoustics for a parallel plate photovoltaic device connected to a potentiometer. A phenomenon of photovoltaic amplification has been formulated.
Keywords: base, power, loudspeaker, potentiometer, photovoltaic amplification, solar energy

1. Introduction

The amplitude of a solar energy wave is defined as the power storage per unit area per unit time. The solar power is stored in a packet of solar energy wave of unit cross sectional area and of length s, the speed of light. Therefore, the solar power intensity I is the product of total power storage capacity for a packet of solar energy wave and the speed of light. The logarithm of two solar power intensities, I_1 and I_2, gives power difference for two solar power intensities. It is mathematically expressed as (Dehra, 2008):

$$\mathrm{Sol} = \log\left(I_1\right)\left(I_2\right)^{-1} \qquad (1)$$

Where, Sol is a dimensionless logarithmic solar noise unit. Mathematically, decisol (dS) is more convenient for solar power systems. Since a decisol (dS) is $1/11^{th}$ unit of a Sol, it is mathematically expressed by the equation :

$$\mathrm{dS} = 11\log\left(I_1\right)\left(I_2\right)^{-1} \qquad (2)$$

The above concepts of acoustics in solar energy are similar to the theory of sound (Rayleigh, 1894). This anticipated theory of acoustics in solar energy is mathematically extensible for determining various acoustical expressions which has been supported by experimental and modelling results for a parallel plate photovoltaic device connected to a potentiometer.

2. Experiments and simulation

The full scale experimental setup for a parallel plate photovoltaic device connected to a potentiometer was installed in an outdoor room facility (Dehra, 2004). The schematic of the experimental setup is illustrated in Fig. 1. An amplifier was built with a pair of glass coated photovoltaic (PV) modules forming a parallel plate duct with a plywood board and connected to a potentiometer. A potentiometer, a wire-wound variable resistor of up to 50 Ω was a wire-wound circular coil with a sliding knob contact. It was used to vary electrical resistance across connected PV modules without interrupting the current. The characteristics of a parallel plate photovoltaic device connected to a potentiometer were established by varying electrical resistance with rotation of knob of a potentiometer. The current-voltage measurements were obtained for determining electric power output with a series electrical circuit connection of a pair of vertically inclined PV modules installed on a wooden frame. The temperatures were measured as a function of volume of a parallel plate photovoltaic device. The non-linear thermal results include measurements of temperatures for PV modules, insulating panel and ventilated air column in the wooden frame. The air velocities were developed in the ventilated air column for transportation of heat both as a measure of buoyancy and fan induced hybrid ventilation. The electrical measurement results of currents, voltages and power with varying electrical resistance of potentiometer are presented in Table 1. The thermal measurement results of temperatures of various components of PV device, ambient air and room air temperatures, air velocities and solar intensities are presented in Table 2. The location and nomenclature of sensors is presented in Table 3. The results of the power output from a potentiometer with rotation of circular knob are illustrated in Fig. 3. A simulation model for prediction of temperature distributions varying with volume of a parallel plate photovoltaic device was developed (Dehra, 2004). The model has been used to predict temperature distributions at pre-defined locations in PV module, plywood board and air flowing through a parallel plate channel. The results of the temperature plots are illustrated from Fig. 3 (a) to Fig. 3 (c).

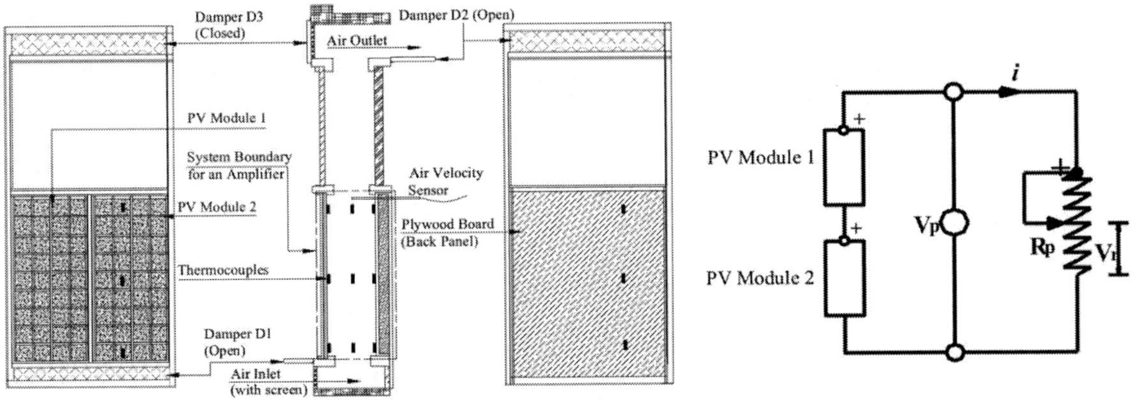

(a) *(b)*

Figure 1. Schematic of experimental setup for a parallel plate photovoltaic device connected to a potentiometer: (a) location of sensors; (b) Electrical circuit diagram

Table 1 Sample electrical measurement results with varying resistance of potentiometer

Rotation	Volts	Amps	Watts	Rotation	Volts	Amps	Watts	Rotation	Volts	Amps	Watts
240°	18.7	-	-	83°	16.3	0.935	15.23	30°	9.7	1.577	15.21
239°	16.5	0.331	5.461	75°	16.0	1.014	16.26	27°	9.0	1.587	14.33
201°	17.4	0.414	7.195	69°	15.8	1.100	17.38	21°	7.1	1.583	11.24
185°	17.5	0.454	7.940	64°	15.5	1.165	18.04	18°	6.2	1.573	9.831
162°	17.3	0.513	8.885	55°	15.0	1.302	19.53	17°	5.7	1.578	9.026
150°	17.18	0.550	9.449	50°	14.5	1.386	20.05	12°	3.9	1.567	6.257
142°	17.19	0.582	10.00	43°	13.2	1.503	19.79	10°	3.2	1.553	4.840
128°	17.1	0.640	10.93	42°	13.1	1.493	19.49	1.5°	0.5	1.593	0.807
107°	16.8	0.750	12.51	37°	11.9	1.536	18.26	1°	0.3	1.59	0.426
89°	16.4	0.884	14.45	32°	10.5	1.567	16.42	0°	-	1.643	-

Table 2 Sample thermal measurement results

Run No.	S (W m^{-2})	E_p (W)	T_o (° C)	T_s (° C)	v (m s^{-1})	$T_p(b)$ (° C)	$T_p(m)$ (° C)	$T_p(t)$ (° C)	$T_b(b)$ (° C)	$T_b(m)$ (° C)	$T_b(t)$ (° C)	$T_a(b)$ (° C)	$T_a(m)$ (° C)	$T_a(t)$ (° C)
1	697.5	31.0	13.5	22.1	0.451	34.5	33.01	36.2	20.2	24.4	27.6	18.7	19.3	22.5
2	725.4	31.1	15.9	22.9	0.362	32.5	33.3	35.7	20.2	23.9	29.1	18.3	19.7	23.3

Table 3 Nomenclature and location of sensors

Locations shown in Fig. 1	$T_p(b)$ (° C)	$T_p(m)$ (° C)	$T_p(t)$ (° C)	$T_b(b)$ (° C)	$T_b(m)$ (° C)	$T_b(t)$ (° C)	$T_a(b)$ (° C)	$T_a(m)$ (° C)	$T_a(t)$ (° C)	Air velocity sensor
y (cm)	15	55	94	15	55	94	15	55	94	99
z (cm)	60	60	60	60	60	60	60	60	60	60
x (mm)	6.2	6.2	6.2	96.2	96.2	96.2	51.2	51.2	51.2	51.2
Latitude of place of experiments	+ 45.495°				Surface azimuth angle			-10°		
Longitude of place of experiments	- 73.5775°				Tilt angle of solar panel			90°		
Geographical location	Montréal, Québec, Concordia				Note: x is horizontal; y is vertical; z is adjacent 3rd axis of x-y plane					

3. Photovoltaic Amplification

The phenomenon of photovoltaic amplification has been observed from the graphs of Figs. (2) and (3). The gain in steady state electrical and thermal functions for a photovoltaic device is a factor of its volume or resistance. This operational characteristic is similar to operation of a loudspeaker. The electrical analog has been used to describe the resonance phenomenon for equivalent mechanical, hydraulic and thermal systems of parallel plate photovoltaic device connected to a potentiometer.

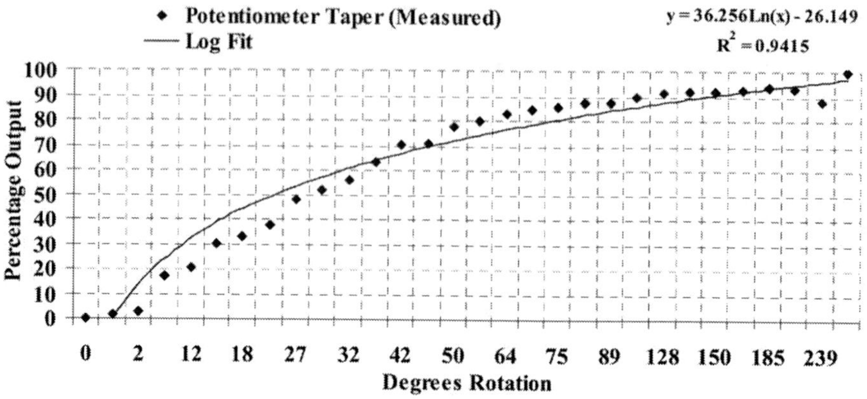

Figure 2. Potentiometer Taper (Measured) with percentage voltage output

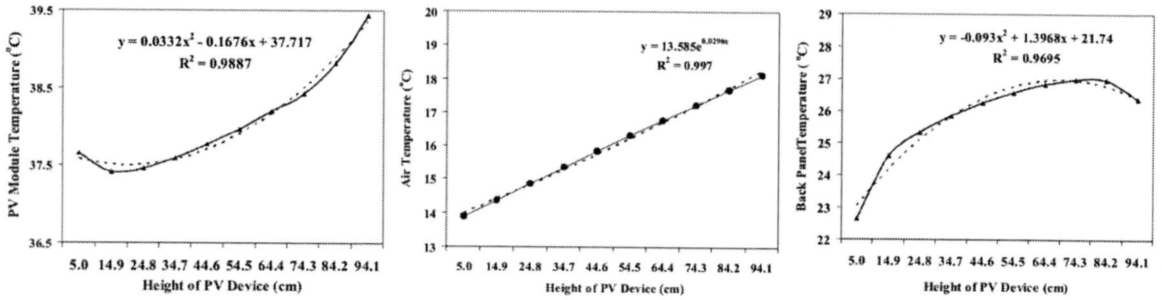

(a) *(b)* *(c)*

Figure 3. Temperature plots with height of PV device: (a) PV module; (b) air; and (c) plywood board

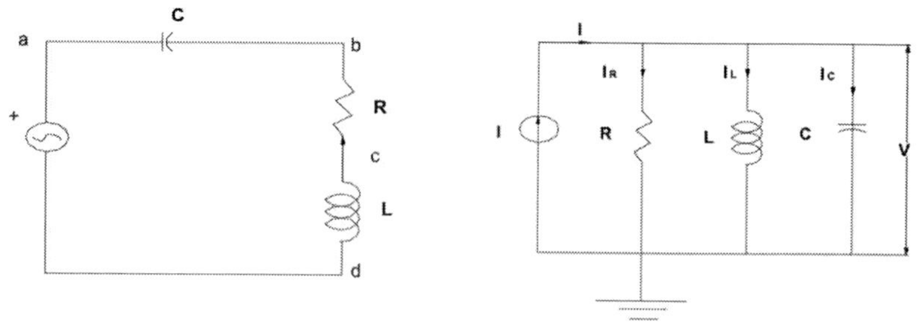

(a) *(b)*

Figure 4. (a) L-C-R series arrangement of resonance; (b) L-C-R parallel arrangement of resonance

3.1 Series resonance

The series and parallel cases of L-C-R arrangement of resonance are illustrated in Figure 4. For the case of series resonance, the equilibrium frequency w_o is written as (Del Toro, 1986):

$$w_o = \frac{1}{\sqrt{LC}} \tag{3}$$

The frequency w_o is called the resonant frequency. Its value is a function of two energy storing elements of the system. At resonance the impedance of the circuit is a minimum, and is equal to R. Consequently, when a series L-C-R circuit is at resonance, the current is a maximum and is also in time phase with the voltage. The power factor is unity. For operation at a frequency below w_o the resultant imaginary part of impedance is capacitive so that the current leads the voltage. When $w>w_o$, the inductive reactance prevails, so that the current then lags the voltage. When the ratio of inductive reactance to resistance in a series L-C-R circuit is high, a very rapid rise of current to the maximum value occurs in the vicinity of the resonant frequency.

3.2 Band Width

The mathematical procedure is used herewith for determining bandwidth at resonance (Del Toro, 1986). The bandwidth is defined as the range of frequency for which the power delivered to R is greater than or equal to $P_R/2$, where P_R is the power delivered to R at resonance. As the magnitude of the current at each half-power point during resonance is same, we can write:

$$i_1^2 R = \frac{1}{2} i_o^2 R = i_2^2 R \tag{4}$$

Where, subscript 1 denotes the lower half-power point and subscript 2 the higher half-power point. The expressions for i_1 and i_2 are then written as:

$$i_1 = i_2 = \frac{i_o}{\sqrt{2}} = 0.707 i_o \tag{5}$$

The bandwidth is defined during resonance as that range of frequency over which the magnitude of the current is equal to or greater than 0.707 of the current at resonance. For $w_oL/R=1/2$, the frequency at the lower half-power point is denoted by w1 and that of the upper half power point is denoted by w_2. Hence the bandwidth is w_2-w_1. Since the current at the half-power points is $io/\sqrt{2}$, it follows that the magnitude of the impedance must be equal to $\sqrt{2}$ R to yield this current. The reactance during resonance is at the lower half-power frequency X_1, is then written as:

$$X_1 = w_1 L - \frac{1}{w_1 C} = -R \tag{6}$$

The minus sign appears on the right side of the equation because below resonance the capacitive reactance exceeds the inductive reactance. Rearranging Eq. (6), leads to:

$$w_1^2 + \frac{R}{L} w_1 - \frac{1}{LC} = 0 \tag{7}$$

The expressions for positive upper half-power and lower half-power frequency are written as:

$$w_1 = \frac{-R}{2L} + \sqrt{\left(\frac{R}{2L}\right)^2 + \frac{1}{LC}} \tag{8}$$

$$w_2 = \frac{R}{2L} + \sqrt{\left(\frac{R}{2L}\right)^2 + \frac{1}{LC}} \tag{9}$$

Therefore the expression for the bandwidth becomes, from Eqs. (8) and (9),

$$w_{bw} = w_2 - w_1 = \frac{R}{L} \tag{10}$$

3.3 Quality factor Q_o of the RLC circuit

By forming the ratio of the resonant frequency to the bandwidth, a quality factor to measure the selectivity or sharpness of tuning of the series RLC circuit is obtained (Del Toro, 1986):

$$Q_o = \frac{w_o}{w_{bw}} \tag{11}$$

When Eq. (10) is inserted into Eq. (11):

$$Q_o = \frac{w_o L}{R} = \frac{X_{LO}}{R} \tag{12}$$

By making use of the knowledge that Q_o is very large in many resonant circuits and systems, the lower and upper half-power points are expressed in terms of the resonant frequency and the bandwidth.
The Eq. (8) is re-written as:

$$w_1 = -\frac{R}{2L} + \sqrt{\left(\frac{R}{2L}\right)^2 + Q_o^2 \frac{R^2}{L^2}} \tag{13}$$

For values of Q_o which are 5 or greater very little error is made by writing:

$$w_1 = -\frac{R}{2L} + \frac{R}{L} Q_o = -\frac{R}{2L} + w_o \tag{14}$$

Inserting Eq. (10) into Eq. (14) yields:

$$w_1 = w_o - \frac{1}{2} w_{bw} \tag{15}$$

Similarly, expression for w_2 is written as:

$$w_2 = w_o + \frac{1}{2} w_{bw} \tag{16}$$

The definition of quality factor is also expressed in terms of power storage. The instantaneous expression for the forced solution of current at resonance is:

$$i_o = I_{mo} \cos\left(w_o t\right) \tag{17}$$

The total power stored by the circuit in both L and C is expressed as:

$$W = \frac{1}{2} L i_o^2 + \frac{1}{2} C V_{CO}^2 \tag{18}$$

Inserting the appropriate expressions for i_o and V_{co} results in:

$$W = \frac{1}{2} L I_{mo}^2 \cos^{(2)}\left(w_o t\right) + \frac{1}{2} L I_{mo}^2 \sin^{(2)}\left(w_o t\right) \tag{19}$$

$$W = \frac{1}{2} L I_{mo}^2 = L I_o^2 \tag{20}$$

Where, I_{mo} is maxima and Io is the rms value of current at resonance. The energy stored at resonance is a constant quantity. From Eq. (12), the expression for Q_o is written as (Del Toro, 1986):

$$Q_o = \frac{w_o L I_o^2}{R I_o^2} = \frac{2\pi f L I_o^2}{R I_o^2} = 2\pi I_o^2 L \left(\frac{I_o^2 R}{f_o}\right)^{-1} \tag{21}$$

It is written in terms of stored energy and dissipated energy as:

Q_o=2πx(total energy stored/energy dissipated per cycle) (22)

It is applicable to any resonant system regardless of its composition.

3.4 Band width for parallel resonance

For the parallel case, the equation for lower half power frequency w_1 is written as:

$$w_1 C - \frac{1}{w_1 L} = -\frac{1}{R} \tag{23}$$

The quadratic equation is written as:

$$w_1^2 + \frac{1}{RC} w_1 - \frac{1}{LC} = 0 \tag{24}$$

Ignoring the negative solution, the desired solution is:

$$w_1 = -\frac{1}{2RC} + \sqrt{\left(\frac{1}{2RC}\right)^2 + \frac{1}{LC}} \tag{25}$$

The expression for the upper half-power frequency is:

$$w_2 = \frac{1}{2RC} + \sqrt{\left(\frac{1}{2RC}\right)^2 + \frac{1}{LC}} \qquad (26)$$

The expression for the bandwidth is:

$$w_{bw} = w_2 - w_1 = \frac{1}{RC} \qquad (27)$$

The expression for quality factor is written as:

$$Q_o = \frac{w_o}{w_{bw}} = w_o R C \qquad (28)$$

4. Discussions

The incident short wavelength electromagnetic waves on a parallel plate photovoltaic device cause development of stresses and oscillations and results in propagation of following composite waves (Dehra, 2007): i) transmission of electricity due to connected external load; ii) the heat waves are propagated at longer wavelength due to exchange of dissipated energy with medium of contact; iii) elastic waves are propagated in photovoltaic modules and polystyrene filled plywood board due to thermal stress development with propagation of heat waves; iv) fluid surface waves are propagated due to combined stress development with heat waves, elastic waves and applied external waves; v) applied external waves are propagated due fan induced pressure and climate-particle oscillations of wind. The sound waves can propagate with combination of fluid surface waves and applied external waves such as in an organ pipe (Dehra, 2007). The sound waves are also acting due to air-borne sound propagation from the surrounding environment. The propagation of acoustic waves is caused due to various stresses and oscillations of a static particle body. The stresses developed due to various forces acting on a cavity wall are classified as: i) fundamental stresses; ii) internal stresses; and iii) external stresses (Dehra, 2007). The fundamental stresses are developed due to presence of gravitational and electromagnetic forces of a solar system. The internal stresses are developed under the influence of fundamental stresses and are defined by properties of a static particle body. The external stresses are developed under an external source of energy such as fan pressure used for active air ventilation through the cavity wall. The oscillations of a cavity wall are assumed to be generated because of stress development with cyclic load of day/night, heating/cooling and compression/expansion applied on a cavity wall. The climate-particle oscillations of wind are also assumed to be acting on a cavity wall. The acoustic waves of cavity wall are propagated due to stresses and oscillations of a cavity wall. The fluctuating forces are generated due to superposition of composite acoustic waves.

The analysis has been presented for series and parallel cases of L-C-R circuit resonance. The cases of resonance can be visualised with the aid of presented experimental and modelling results. Inductance is mass for the mechanical system for propagation of elastic waves. Capacitance is heat storage capacities of photovoltaic modules, air and polystyrene filled plywood board. Polystyrene filled plywood board is vulnerable to fire as soon as heat waves propagated with frequency matching with its latent heat of vaporisation is reached. Resistance is developed due to thermal and fluid resistance in energy storage elements of photovoltaic modules, air and polystyrene filled plywood board. The series case of L-C-R resonance occurs with propagation of elastic waves of mechanical system of photovoltaic cavity wall. The parallel case of L-C-R resonance occurs with heat waves (C-R) and fluid surface waves (C-R) in conjunction with inductance (L) due to mass of photovoltaic cavity wall and resistance (R) due to temperatures of ground surface and ambient air.

5. Conclusions

A theory of acoustics in solar energy has been presented. The sample measurements and modelling results are presented for a parallel plate photovoltaic device connected to a potentiometer. A phenomenon of photovoltaic amplification has been formulated similar to resonance conditions applicable to the theory of sound. The resonance conditions are presented for series and parallel cases of L-C-R connected elements. The maximum energy transfer can occur during resonance with transmission of light, sound, noise, heat, electricity, fluid and fire from a parallel plate photovoltaic device connected to a potentiometer (Dehra, 2007). The expressions for bandwidths and quality factors during occurrence of resonance at maximum energy transfer are presented.

References

Rayleigh, J. W. S., 1894. The Theory of Sound Vol. 1, New York, USA, Concordia: Dover Publications.

Dehra, H, 2004. A numerical and experimental study for generation of electric and thermal power with photovoltaic modules embedded in building façade, Ph.D. thesis, Department of Building Engineering, Concordia University, Montréal, Québec, Concordia, August 2004.

Dehra, H., 2007. A unified theory for stresses and oscillations, Canadian Acoustics, vol. 35 (3), pp. 132-133.

Dehra, H., 2008. The noise scales and their units, Canadian Acoustics, vol. 36 (3), pp. 78-79.

Del Toro, V., 1986. Electrical Engineering Fundamentals, New Delhi, Sindhudesh, Concordia: Prentice-Hall.

(31f) Controller Performance Monitoring & Control Improvement - the Step Before Advanced Control

John Gerry and George Buckbee, ExperTune Inc, Hartland, WI

Advanced Process Control projects that use Model Predictive or other sophisticated control algorithms depend on:

1) Optimized and working regulatory controls - "so the PID controller will do what the MPC controller tells it to do"

2) Choosing true interacting loops - to make the most cost effective use of the expensive Advanced control algorithms

This paper explains how both of these goals can be met cost effectively by in performance monitoring software. The first step is using plant-wide correlation analysis to find the control loops with high degree of interaction. This to help narrow the field of manipulated and controlled variables for Advanced Control. This presentation also explains how a performance monitor ensures the PID controllers, valves and sensors are healthy, delivering good base level control for the Advanced controllers to use. The use and using the recommendations of performance monitoring ensures the Advanced controls have a solid foundation to optimize the plant.

(47c) Compact Heat Exchange Reactor for In Process Improvement

Zhijun Jia[1], **Steven J. Vallee**[1], Burtron H. Davis[2] and Dennis Sparks[2], (1)Chart Energy & Chemicals, La Crosse, WI, (2)Center for applied energy research, University of Kentucky, Lexington, KY

Chart has developed a next generation reactor that is very suitable for highly exothermic or endothermic reactions such as Fischer-Tropsch (FT) synthesis or Steam Methane Reforming (SMR). Chart's Compact Heat Exchange Reactor (CHER) is an advanced technology which combines heat exchanger, mixer, and reactor together in a compact structure. The reactor consists of multiple reaction and cooling layers alternately arranged. Each reaction layer contains multiple mini-channels each in the mm scale size range.

A study of FT synthesis using Chart's CHER was performed. Mini-channels in this reactor were packed with catalysts provided by Burt Davis's group at Kentucky University. The reactor was continuously running for three weeks at operating conditions of 220C and 15 bars. The tested reactor was made of aluminum which provides an excellent heat transfer characteristics. The results indicated that the compact heat exchange reactor provided unique features to improve the process efficiency and productivity. One of the significant benefits by using CHER in FT or SMR processes is improved temperature control which allows the reactor to operate efficiently at higher conversions than traditional tubular reactors.

Paper 47b

PROCESS INTENSIFICATION OF MULTI-EFFECT EVAPORATION AND CRYSTALLIZATION PROCESSES

C. B. Panchal ** and Vyjayanthi Alagharu
E3Tec Service, LLC
Clarksville, MD 21029-1203
cpanchal@msn.com

Emmanuel Dada
ChemProcess Technologies, LLC
League City, TX 77573
emmanuel.dada@ymail.com

Chris Thomas
FMC Alkali Chemicals
Green River, WY 82935
Chris.C.Thomas@fmc.com

** To whom correspondence should be sent

Key Words: multi-effect, evaporation, crystallization, enhanced tubes, process intensification

Prepared for Presentation at the 2011 National Meeting, Chicago, IL March 14-17, 2011

Unpublished

PROCESS INTENSIFICATION OF MULTI-EFFECT EVAPORATION AND CRYSTALLIZATION PROCESSES

C.B. Panchal[a], Vyjayanthi Alagharu[a], Emmanuel Dada[b] and Chris Thomas[c]
[a] E3Tec Service LLC, Clarksville, MD 21029
[b] ChemProcess Technologies, LLC, League City, TX 77573
[c] FMC Alkali Chemicals, Green River, WY 82935

Abstract

Multi-effect evaporation and crystallization are key separation processes for concentrating and purifying inorganic and organic solid products. They are second only next to distillation in their consumption of energy with estimated energy consumption of greater than 660 trillion Btu/yr [ERDA, 1976], or the equivalent of 110 million barrels of oil per year. A small improvement in the energy efficiency can have major impact on overall energy savings and reducing the C-Footprint. The objective of this DOE SBIR Phase I project focused on two specific aims: 1) to evaluate enhanced-tube performance for process intensification; and 2) to characterize fouling propensity of enhanced tubes in multi-effect evaporation and crystallization. The process analysis was performed for the triple-effect and mechanical vapor recompression (MVR) crystallizers of the soda ash process. Once a benchmark analysis was performed, the impact of enhanced tubes was evaluated in terms of reduced power in MVR and reducing the primary-steam pressure in the triple effect crystallizer. The potential of energy efficiency in the MVR is direct and significant. The primary steam in triple-effect is generally derived from power generation; therefore, by reducing the steam pressure increases the power generation. Optimally designed enhanced tubes have shown to reduce fouling propensity in other applications. The interactive effects of wall temperature, wall shear stress, and secondary flows in the vicinity of the heated wall surface are the governing factors associated with fouling. Two types of enhanced tube were evaluated for applications in soda ash process. Applications of enhanced tubes in vertical or horizontal forced convective evaporator as well as falling-film evaporators were also evaluated.

Introduction

Process industries with heavy applications of multi-effect evaporation and crystallization include soda ash and caustic soda manufacturing, salt production, purification and separation of organics, black liquor concentration in the pulping process, recovery of distillers' dried grains with solubles (DDGS) in ethanol corn mills, production of ethylene and propylene glycol, agricultural chemicals, sugar production, fruit juice concentration, and various environmental treatments. The market potential of improved energy efficiency in multi-effect evaporation and crystallization is high. The production of soda ash and caustic soda are listed 9th and 10th in the US top 50 chemicals at 11,838 and 11,371 KTA (thousand tons per annum) [Energetics, 2000], and evaporated salt production is 4.3 KTA. Pulp, paper and paperboard mills account for 95% of energy use in the US paper and allied products industries, and 12% of total manufacturing energy use in the U.S [Nilsson et al., 1995]. Production of fuel ethanol is rapidly increasing in the U.S. to meet the national goal. Production is expected to increase from the current level of about 6 billion gallons per year to 36 billion gallons per year by 2020 [EERE, 2007], most of this production is expected to be based on corn and cellulosic feedstock. The net-energy value of corn-based ethanol can be achieved only when the energy credit for co-products, which requires multi-effect evaporators, is included [Shapouri et al.,

2002]. Therefore, it is essential to develop an energy-efficient evaporation process for product recovery.

Currently enhanced tubes are not used in multi-effect evaporators and crystallizers and monitoring-based fouling mitigation is not practiced. As a first impression, qualifying enhanced tubes in multi-effect evaporators and crystallizers seems an ordinary step and the industry should be able to pursue it without further developments. Response to this perception lies in understanding the complex multi-phase solid/liquid/vapor flows in evaporators and crystallizers. The industry is not likely to implement new enhanced tube geometry without validated thermal performance, effective and predictive fouling mitigation, and clearly defined economics in terms of Return on Investment (ROI).

Objectives and Specific Aims

The major objective of this SBIR project was to make fundamental changes in designing heat transfer equipment in multi-effect and MVR evaporation and crystallization processes that would have significant impact on process intensification; energy efficiency and productivity. In order to establish the applicability of enhanced tubes, this SBIR project focused on the following specific aims.

a. Optimal design of enhanced tubes in multi-effect and MVR evaporators and crystallizers;
b. Characterization of fouling propensity of enhanced tubes; and
c. Development of an innovative monitoring-based fouling mitigation system.

Application of Enhanced Tubes

This project focused on two types of enhancements; 1) Twisted Tubes®; and 2) spirally indented tubes with micro-fins on outside. They are ideally suited for single-phase flow, thin-film evaporation (rising or falling film) and condensation heat transfer in multi-effect evaporators and crystallizers. Twisted Tubes® (Figure 1) induce swirling flows that increases the thermal performance by more than 30% in single-phase flow and greater in two-phase flows, without corresponding increase in pressure drop. Spirally indented tubes, shown in Figure 2 enhance the thermal performance by interrupting the boundary layer and producing turbulent eddies in the vicinity of the wall without significant flow recirculation zones.

Figure 1 Enhanced Twisted Tubes® (Courtesy of Koch Heat Transfer Co.)

Selected tube inserts used in other process industries were also evaluated. A careful analysis of application of commercially available tube inserts showed that spirally twisted tapes or ribs and flow-induced oscillating wire inserts are viable candidates. Tube inserts can be installed without removing the tube bundle; therefore, it is a low-cost option for enhancing the thermal performance and reducing the fouling propensity. However, it is essential to optimize the geometry of tube inserts for multi-effect evaporator and crystallizer heat exchangers. The oscillating

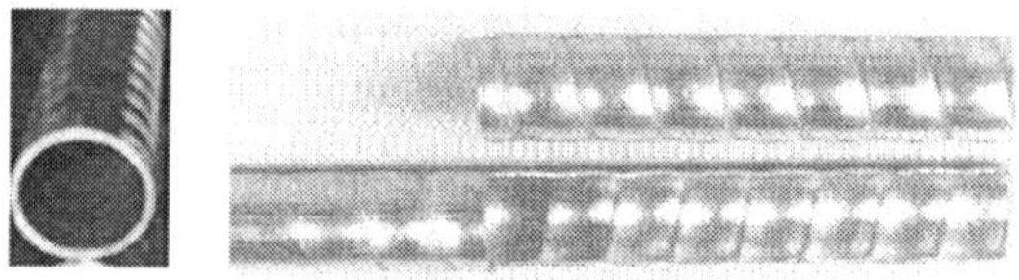

Figure 2 Spirally indented tubes

wire has shown to reduce fouling by generating secondary flows in the vicinity of the tube wall as well as disturbing the deposition mechanisms due to the oscillating movement of the wire. The major disadvantage of tubes inserts is that they cannot be readily removed for mechanical cleaning of heat exchanger tubes. Therefore, cleaning-in-place (CIP) methods are required for these inserts.

The major categories of heat-transfer equipment used in multi-effect evaporation and crystallization processes are represented by schematics in Figures 3 through 6. The caldaria evaporative crystallizer is generally used for salt production, granular compounds and sugar. Recirculation of solution is mostly caused by the thermo-siphon effects of two-phase flow inside tubes with added effects of recirculation impeller located in the center. Based on proven performance of twisted tube in vertical thermo-siphon reboilers in distillation processes [Panchal and Ljubicic, 2008], optimally sized *Twisted Tubes®* are ideally suited for calandria evaporative crystallizer.

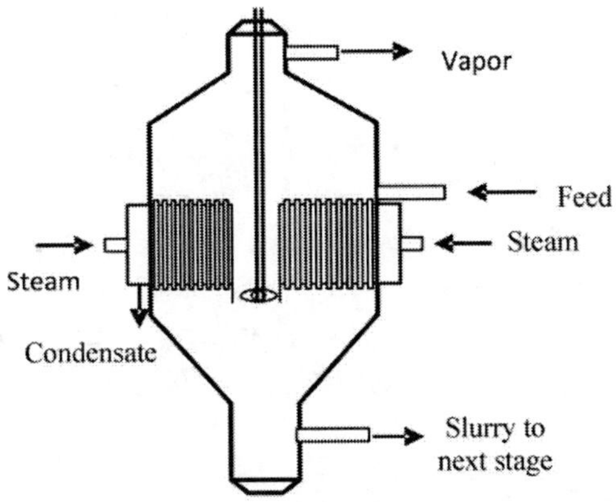

Figure 3 Calandria evaporative crystallizer

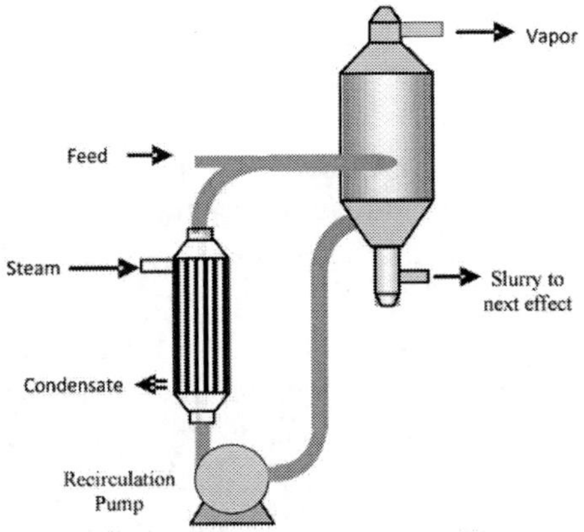

Figure 4 Forced-circulation evaporative crystallizer

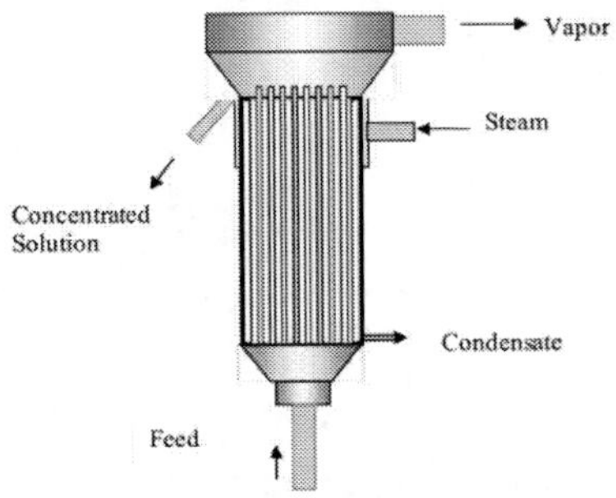

Figure 5 Vertical Rising-Film Evaporation

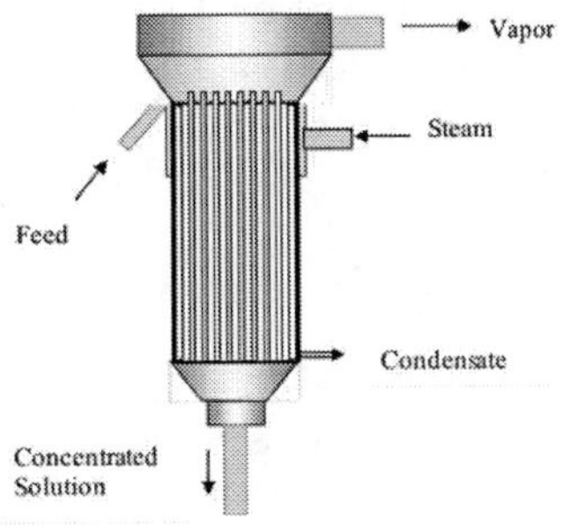

Figure 6 Vertical Falling-Film Evaporation

Forced-circulation evaporative crystallizers are widely used in various industrial processes, Figure 4. Evaporation can occur in the heat exchanger that would concentrate the solution. Alternatively, the solution is super-heated under static pressure, and the solution is allowed to flash producing low-pressure steam and crystal in the crystallizer. *Twisted Tubes®* are ideally suited for both modes of heat transfer; single-phase as well as phase-change. However, optimal size (twisting angle and pitch) would be different for single-phase and two-phase heat transfer modes. *Spirally indented tubes* should be considered for single-phase heat transfer only. Thin-film evaporation is commonly used for concentrating the solution without crystallization, such as caustic soda, food products, and in the paper pulping processes. Besides high thermal performance, thin film evaporation maintains the film temperature within a specified range for product quality and minimizing fouling deposition that can be caused by side reaction (specifically for food products) and crystallization (due to the presence of no-process elements) at higher wall temperatures. Twisted tubes should be considered for thin-film evaporation because swirling flow will maintain effective distribution of thin film on the wall; minimizing hot spots. In the MVR process, shown in Figure 7, vapor generated from the crystallizer or evaporator is compressed, which becomes the heat source. The enhanced thermal performance lowers the temperature difference between the compressed steam and the process fluid and thereby reduces the power consumption.

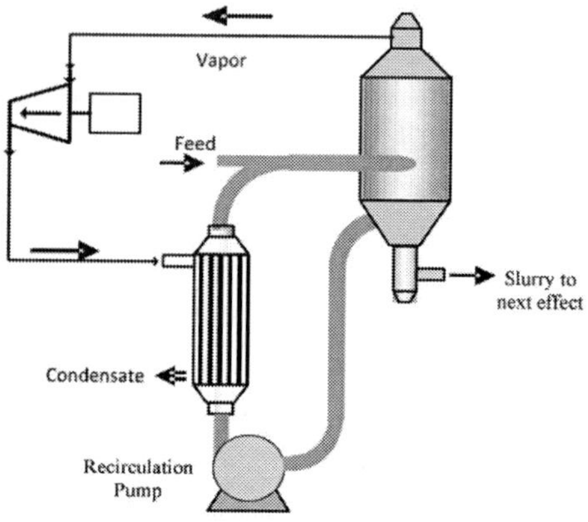

Figure 7 MVR Crystallizer/Evaporator

Steam Economy Analysis of Triple-Effect Soda Ash Process

A typical triple-effect soda ash process is shown in Figure 8. Primary steam is used in the 1st effect and steam generated in 1st and 2nd effects is used in the next effect as heat source. The soda ash solution is circulated at high flow rates to maintain temperature rise < 5 °C. The thermal performance of heat exchanger in each effect governs the temperature difference between the heat source (i.e. steam pressure) and process fluid. Poor thermal performance of heat exchangers would raise steam pressures in each stage, and thereby increase pressure of the primary steam. Source of the primary steam is generally the last stage of a power plant. Increase in steam pressure would adversely impact the power generation. According to EPRI guideline, 0.4" HgA increase in back-pressure of steam due to fouling of steam condenser corresponds to power penalty of about 0.5%.

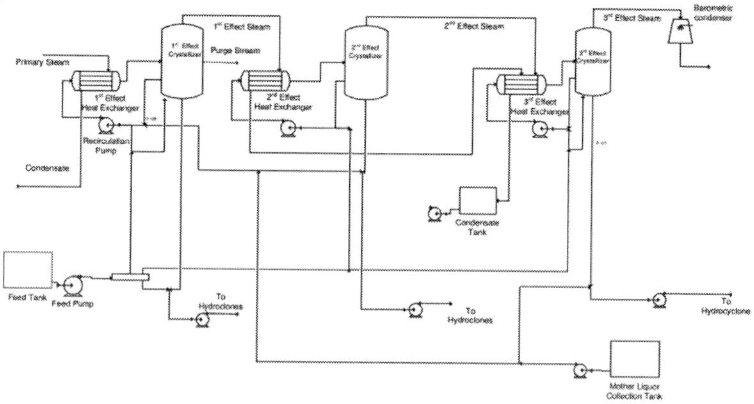

Figure 8 Process flow diagram of the soda ash triple-effect

The benchmark analysis performed by Aspen+® analysis followed by parametric analysis using SimuLink®. The analysis shows the benchmark steam economy is about 2.34. The steam economy could be even lower in normal operation due to poor thermal performance and fouling of heat transfer equipment. Log-mean temperature differences (LMTD) of the evaporative crystallizers vary between 10°C and 25°C, depending on recirculation rates, fouling, and the effectiveness of removal of non-condensable gases. By enhancing the thermal performance and reducing the fouling propensity of the three heat exchangers, LMTDs would be reduced with resulting effects on lowering the pressure of the primary steam from 257 kPa to a lower value of 152 kPa. This reduction in steam pressure would increase the power generation by 2% to 3%.

Performance Analysis of MVR

A typical MVR process diagram for soda ash is shown Figures 9. Aspen+® analysis was performed to establish the benchmark performance followed by SimuLink® analysis for parametric analysis. In addition, an Excel-based detailed heat transfer performance analysis was performed to determine an optimum enhancement. The benchmark power consumption of this case study is 69.7 kWh/metric ton (MT). The results show that by enhancing the heat transfer coefficient by 40%, which can be achieved by enhanced tubes described above, the power consumption can be reduced by about 9% by lowering the compression ratio. Further improvements would be achieved by reducing the effective fouling resistance and the effect of non-condensable gases. A conservative estimate of lowering the combined effects of fouling and non-condensable gases by one-half was used in this analysis. The resulting effects of heat transfer enhancement and lower effective fouling resistance are quite significant. The compressor power is reduced by 24% compared to benchmark power.

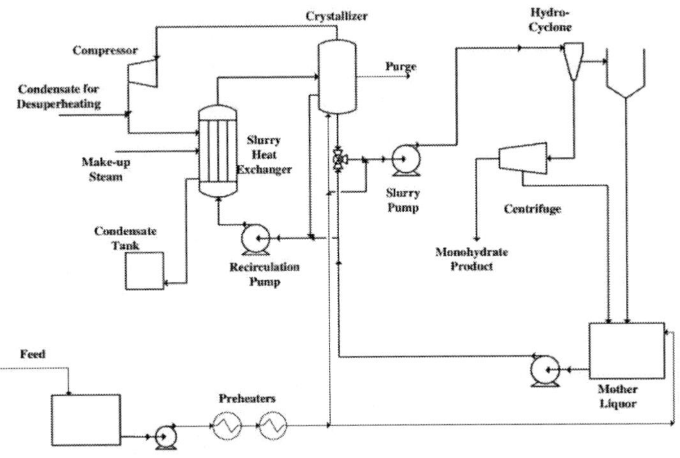

Figure 9 Process flow diagram of soda ash MVR

Fouling Characterization of Enhanced Tubes

In order to characterize fouling of enhanced tubes, it is important to understand the fouling mechanisms and interactive effects of governing mechanisms of precursor formation and subsequent physical processes associated with fouling deposition. Crystallization and particulate deposition are the two dominant fouling mechanisms in multi-effect evaporation and crystallization processes. Grace [1975] provided a survey of evaporator scaling of alkaline pulp industry, where crystallization fouling due to complex formation by non-process elements is a major concern. The fouling problem results from co-deposition of sodium carbonate (Na_2CO_3) and sodium sulfate (Na_2SO_4) [Schmidl and Frederick, 1998]. *With increased use of recycled water for achieving zero discharge, fouling caused by non-process elements are increasingly becoming a major concern in many industrial processes.*

Fouling in multi-effect evaporators and crystallizers can be modeled in terms of wall shear stress, wall temperature, and chemical/physical characteristics of foulant as shown by equation below developed for plain tubes.

$$\frac{dR_f}{dt} = R_f^* \left\{1 - \exp\left(-\frac{C\tau\lambda_f t}{\varphi}\right)\right\}$$

This equation is based on the assumption that foulant can be removed from the surface by wall shear stress. For crystallization fouling, increasing wall shear stress effectively enhances the heat transfer coefficient and hence the thickness of the thermal boundary layer and lowers the wall temperature closer to the process fluid temperature. The combined effects significantly reduce the fouling propensity. However, achieving higher wall shear stress, just by increasing the fluid velocity, would increase pressure drop and hence pumping power. With better understanding of the fouling mechanisms and fluid dynamics in the vicinity of enhanced tubes, the investigators can confidently conclude that by creating secondary flows would effectively reduce the fouling propensity. This is the basic concept of this SBIR project. Enhanced tubes create swirling flows in the vicinity of the wall surface that would have high effectiveness to reduce the fouling rate as expressed in equation above. Optimally designed spirally indented tubes would also induce secondary flows at micro scale in the vicinity of the wall, in addition to reducing the effective thermal boundary layer due to periodic interruptions. Panchal and research associates have developed the following threshold fouling correlation for organic fluid fouling involving chemical reaction in the boundary layer [Panchal et al. 1997]. This correlation is being used to evaluate threshold fouling conditions and design heat exchangers as well as characterizing fouling of enhanced tubes. These two fouling correlations can be integrated to develop a threshold fouling correlation for enhanced tube in evaporation and crystallization processes.

$$\frac{dR_f}{dt} = \alpha Re^{\beta} Pr^{-0.333} e^{(-\frac{E}{RT})} - \gamma\tau$$

The influence of enhanced surfaces on transport processes, shear stress distribution, secondary flows in the vicinity of the heated surface, and wall temperature alter formation of fouling precursor and subsequent deposition. Crystallization fouling in evaporative crystallizers occurs in the thermal boundary layer. Therefore, the distribution of shear stress, secondary flows, and wall temperature have direct impact on the fouling characteristics of enhanced surfaces. The literature analysis clearly demonstrates advantageous characteristics of optimally enhanced tubes for mitigating fouling in single-phase as well as two-phase flow conditions. However, limited data are available to determine fouling characteristics of enhanced tubes in multi-effect evaporation and crystallization with free-flowing solids, which should be pursued using field-thermal monitoring unit as has been done for petroleum fouling [Panchal et al., 2004].

Conclusions

This SBIR project has shown ***techno-economic feasibility*** of ***enhanced tubes*** for applications in multi-effect evaporation and crystallization processes with potential of significant process intensification. The specific conclusions are as follows.

- Twisted Tubes® and spirally indented tubes, with or without micro-fins on outside for enhancing steam condensation, are potential candidates for multi-effect evaporation and crystallization processes.
- Aspen+® benchmark analysis combined with Excel-based thermal performance and SimuLink®-based parametric analyses provide effective tools to evaluate process intensification in multi-effect and MVR evaporation and crystallization processes.
- Improving the thermal performance of heat transfer equipment in the MVR process has high potential rate of return.
- Fouling monitoring units designed for the petroleum industry can be modified for multi-effect evaporation and crystallization processes.

Acknowledgement

This work was supported by the U.S. Department of Energy, the Office of Basic Science, Small Business Innovative Research (SBIR) Program.

References

ERDA, 1976, "Energy Conservation: Evaporation – A Prime Target for Energy Conservation," Energy Research and Development Administration Report, COO/2870-1.

Energetics, 2000, "Energy and Environmental Profiles of the U.S. Chemical Industry," Report prepared for the U.S. Department of Energy, Office of Industrial Technologies.

EERE, 2007, "Growing America's Energy Future," *The U.S. Department of Energy, EERE/ITP.*

Grace TM, 1975, "Survey of Evaporator Scaling of Alkaline Pulp Industry," Project 3234, Report 1, The Institute of Paper Chemistry, Appleton, WI.

Nilsson LJ, Larson ED, Gilbreath KR, and Gupta A, 1995, *Energy Efficiency and the Pulp & Paper Industry*, American Council for an Energy-Efficient Economy, Washington DC.

Panchal CB, Malone P, Stomeirowski S, and Cao T, "Mitigation of crude oil fouling by tube inserts and enhanced tubes," *Proceedings of the AIChE Spring Meeting*, New Orleans, LA, April 2004.

Panchal CB, Kuru WC, Liao CF, Ebert WA, and Palen JW, "Threshold conditions for crude oil fouling," *Understanding Heat Exchanger Fouling and Its Mitigation, Proceedings of the 1997 Engineering Foundation Conference*, Begell House, Inc., New York, NY 1997.

Panchal, CB, Malone, P., Stomeirowski, S., and Cao, T, 2004, "*Preheat Train Fouling: Effects of Process Parameters and Mitigation by Tube Inserts and Enhanced Tubes*," Proceedings of the AIChE Spring Meeting, New Orleans, La.

Panchal CB and Ljubicic B, 2008, "Design Approach for Mitigating Fouling of Reboilers and Feed/Effluent Heat Exchangers," Proceedings of the AIChE Meeting, New Orleans, LA.

Schmidl W, Frederick W; 1998, "Current Trends in Evaporator Fouling", Proceedings of International Conference on Chemical Recovery, TAPPI Press, Atlanta, GA.

Shapouri H, Duffield JA, Wang M, 2002, *The Energy Balance of Corn Ethano*l, The US Department of Agriculture Report No. 813.

Lonza MicroReactors – Scale-up

EHRFELD Mikrotechnik BTS
A Bayer Technology Services Company

Preclinical | Clinical Phase I - III | Commercial Production

Lab Reactor	Plate Reactors A6 & A5	Plate Reactor A4
	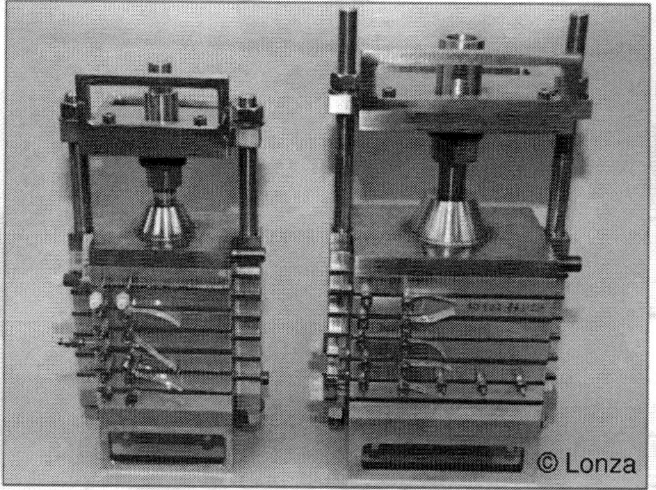	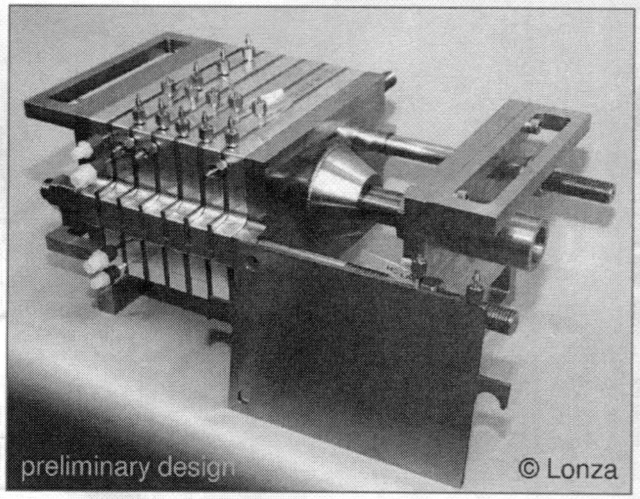
1-50 g/min	30-150 & 100-300 g/min	200-600 g/min

Ehrfeld Microreactor Systems

EHRFELD Mikrotechnik BTS
A Bayer Technology Services Company

Pilot- and production scale apparatus for chemical industry

Modular MicroReaction Systems for laboratory use

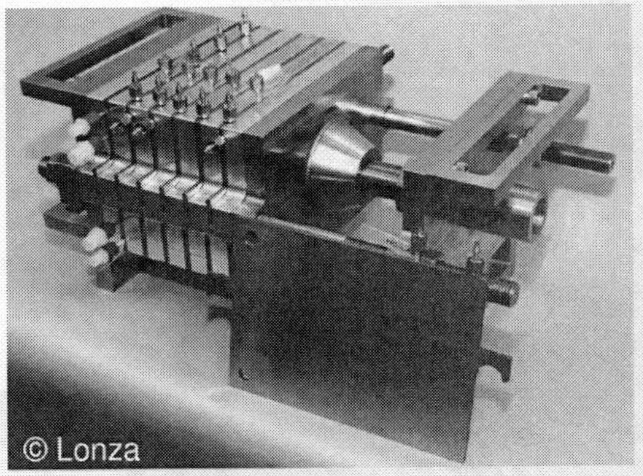

Reactor systems for pharmaceutical industry

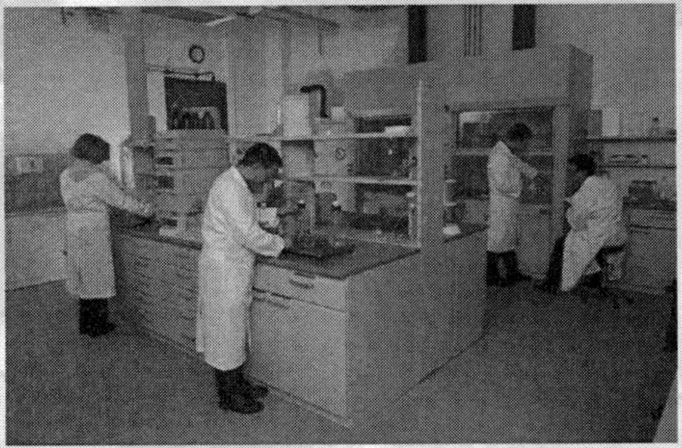

Feasibility studies, training, contract development

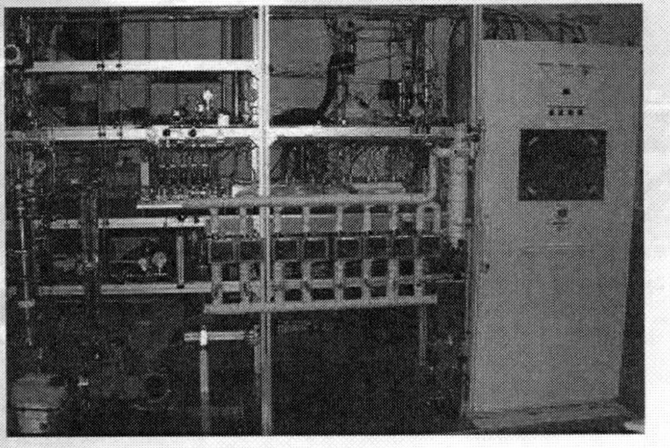

Customized plants

(47d) Membrane Extraction (pertraction) as a Tool for Process Intensification

Ikenna S. Ngene, Separation Technology, TNO Science and Industry, 2600 JA/Delft, Netherlands and Mark Roelands, Process Intensification, TNO Science and Industry, 2600 JA/Delft, Netherlands

Classical extraction is a well known technique used in the separation of components within the industry. Pertraction was developed as an extraction process which reduces the effect of emulsion formation resulting from mixing two solvent phases. In pertraction, a membrane is used to separate the two phases. Typically, the solute of interest is present in the aqueous phase and is transported into an organic phase across a hydrophobic membrane. Considering the independent mobility of the two phases, there is better control over optimization. With the use of fibers, the contact area for these pertraction processes can be significantly increased.

In terms of process intensification, there is presently a drive from batch to continuous processing in fine-chemicals industries. This drive offers advantages with respect to volumetric productivity, yield and selectivity, footprint and safety. Pilot studies of micro-structured reactors in a production environment are in progress. It makes, however, little sense to shift only the reactor to continuous operation and not the work up which would result in accumulation of the reaction mixture in storage vessels. The next logical step is the introduction of continuous intensified separation equipment that has a similar performance with respect to separation efficiency as continuous reactors regarding productivity.

This study is focused on the use of a flat sheet membrane contactor for pertraction with two different chemical applications tested. In the first part, an aqueous work up of a synthetic organic reaction mixture was carried out. This work up was aimed at the removal of spent organic acids and bases from the reaction mixture using pertraction. The work up was successfully carried out without emulsion formation and potentially with a smaller volume of washing water. The second application studied was the dehydration of a reaction mixture, with the objective of removing both dissolved and dispersed water formed. In the latter case the test system was an esterification reaction and dehydration into a hygroscopic solution. For both applications the mass transfer coefficient was determined in order to calculate the surface area of membrane required for an industrial module.

New Optimization-Based Approach to Process Synthesis – towards the Full Integration of Process Design, Operation and Control

Nikola M. Nikačević [1], Adrie E.M. Huesman [1], Paul M.J. Van den Hof [1], Andrzej I. Stankiewicz [2]

[1] *Delft Center for Systems and Control, Delft University of Technology, Delft, The Netherlands,* n.nikacevic@tudelft.nl
[2] *Process & Energy Department, Delft University of Technology, Delft, The Netherlands*

Conceptual process design still follows a traditional procedure which consists of three sequential stages, as depicted in Figure 1. In the first stage, the operation mode is selected according to the particular process and manufacturing demands. The decision is strictly based on heuristic rules, for instance batch operation is preferable for small capacities and multi-products while continuous operation is recommended for bulk production of a single product. In the second stage, the process is designed according to the process' physical and chemical characteristics, desirable production rate and quality and related economical issues. Although in this step equipment dimensions are commonly calculated using mathematical models, this phase also heavily relies on heuristics. In most cases predefined process units for a specific application are used and a selection of units is completed according to this practical method. In the third phase, a control system is designed for the purpose of maintaining the specified product quality and quantity, as well as process safety. This procedure is well established in practice; it offers ready-made mathematical models and technical solutions. However, it does not allow the interaction of design, operation and control and therefore it is generally suboptimal from both economical and environmental perspectives.

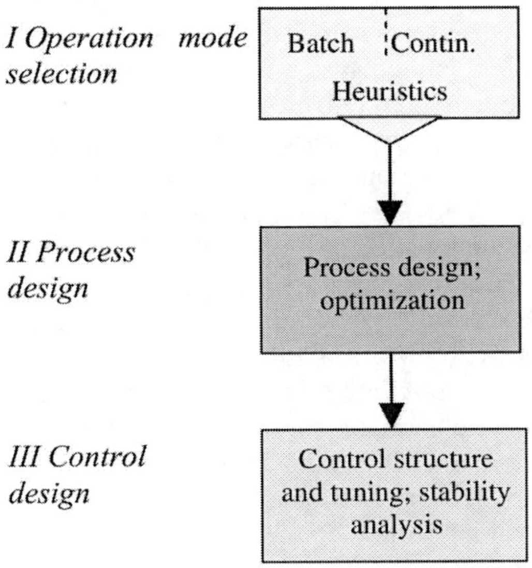

Figure 1. Traditional process synthesis

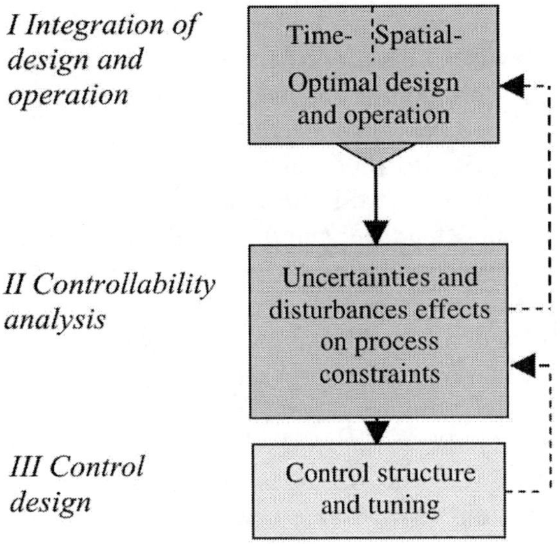

Figure 2. New optimization-based process synthesis

A new approach to chemical process synthesis is going to be presented and illustrated on an example. The concept is based on dynamic optimization and it also contains three stages as presented in Figure 2. In contrast to traditional approach this concept provides strong interactions between process operation, design and control. In the first stage, the complete integration of process design and operation should be attained by means of dynamic optimization, based on simple first principles mathematical models. Optimization objectives are based on economical modeling and they may include environmental issues. Optimization results should provide optimal process design and optimal operation. Proposed process synthesis concept is it is out of the limits of traditional unit operation approach and it is determined by exploration of phenomena and manipulation of fluxes and driving forces. It should be emphasized that the operation mode is not fixed *a priori* (as in Fig. 1) but is a result of the optimization. Therefore, this concept promotes the exploitation of process intensification principles and methods for process design. It also explores the possibilities for actuation improvement for optimal operation and control.

In the second stage (see Figure 2) controllability study for the optimal process solution (achieved in the first phase) is to be performed, by means of robust optimization techniques. In this analysis, the effects of model uncertainties, specifically physical parameters uncertainties, and process disturbances on the process constraints are to be examined. This analysis may lead to adjustment of actuation variables, in order to avoid violation of the constraints. In the third phase, the control system is conceptually designed, taking the controllability results into account. Its target is to follow the optimal operation path as effectively and robustly as possible. It is reasonable to assume that existing advanced control systems, like (nonlinear) model predictive control, or optimal control, will function adequately for the optimal designs and realization of the optimal operation. A closed loop information is transferred back to the stage two (dashed line) in order to perform closed-loop controllability analysis, which could eventually lead to additional actuation alteration (dashed line to the first stage). In this presentation, the first and most important stage of new process synthesis concept will be addressed, while the second stage will be tackled in the following paper.

The methodology for attaining the integration of process operation and design (stage one in Figure 2), could be generalized as follows. First, after a chemical problem is given (with known reaction mechanism and properties) an economic objective function is modeled, which includes the revenues for a production of desired product and costs for: a) raw materials b) investments, c) energy, d) environmental issues. The problem is then systematically analyzed to consider as much as (physically) possible structural and operational enhancements, in order to maximize fluxes and driving forces within the process. For that reasons various process intensification methods are considered. Formulation of a single unique mathematical model which incorporates different process structure options and ways of operation is not attainable at the present. Therefore, the proposed method implies the decomposition of a generic problem into two or more sub-problems to be subjected to an optimization (Figure 2). One type of general model will generate more spatially-oriented solutions, operated often in continuous regime, both steady and unsteady state. The second form of model will result in more timely-oriented solutions, operated more frequently in discontinues or semi-continuous mode. After the concurring problems are selected and models are set, all degrees of freedom (control variables) need to be listed and classified between time-variable or time-invariant. Apparent physical constrains are to be incorporated in the optimization set-ups, as usual. After dynamic optimizations (MIDO) are performed, the target function values will be compared for a selection of the more beneficial solution, which would further be treated in stages two and three.

Proposed process synthesis methodology, precisely its first phase, is illustrated on an industrially relevant example – methane steam reforming (MSR) accompanied with water gas shift reaction (WGS). The goals are to have: a) high conversion of methane; b) efficient separation of hydrogen; c) high driving forces for reaction and separation; d) low energy costs; e) low capital costs and f) compact design. To overcome the equilibrium limitations and achieve high conversion of methane, a functional integration of catalytic reactor with H_2 selective membrane is a very good process intensification option. A membrane will also provide immediate separation of a product from the undesired product and allow the use of lower temperatures. The alternative could be an integration of catalytic reaction with *in situ* adsorption; however this option was not consider at present, in order to have manageable and illustrative example. To attain high driving forces for the membrane separation, as well as for the reactions, consecutive order of process functions in one unit (reaction-separation-reaction-...) is considered, aside to fully integrated system. Without going into technical details, the spatially-distributed continuous systems could be realized as a tubular catalyst bed reactor containing a number of membrane-wall tubes for the H_2 permeate, if the process functions are integrated. The other, time-distributed class of systems could be designed as a rotating bed reactor with membrane-wall tubes. To minimize diffusion resistances, a structural packing with Ni catalyst coating will be assumed. It should have a large surface area, but high bed void fraction, similarly like foam type of packing.

Alongside mentioned structural and functional intensification, the actuation improvements should be examined systematically and rigorously within the optimization. The first reaction (MSR) is highly endothermic and requires high temperatures and the second reaction (WGS) is slightly exothermic and requires moderate temperatures. In order to attain optimal reaction temperature profile a spatially distributed heat supply will be considered and optimized. The optimal heat supply, with high energy efficiency could possibly be realized by using electric current as a direct heat sources in the catalyst. Alternatively, microwaves could be used for efficient and selective heat supply to catalyst. Further, optimal control in space and time with these alternative energy sources could be achieved effectively due to fast dynamic characteristics electromagnetic fields. For the continuous system, unsteady state (periodic) operation is also examined, as another possibility for operational improvement.

Economical target function comprises a maximal production of hydrogen with a minimal costs for catalytic and membrane surfaces and energy consumption. The dynamic models are simplified using common assumptions in order to attain convergence during the optimizations with a number of constraints and control variables. Altogether, six optimization problems have been derived and analyzed. For all analyzed systems the dynamic optimization results provide optimal design parameters and optimal operation, satisfying process constraints. Optimal design parameters include: number of parallel systems, number of membrane tubes per unit, diameters of units and membrane tubes, lengths, diameter of foam pore, etc. Optimal operation parameters are: inlet steam to methane ratio, inlet temperature and pressure, inlet flowrate of sweep gas, spatially distributed heat supply, etc. The base case system for the comparison of economic performance consists of one reaction unit followed by one separate membrane unit, similarly to a conventional system. Finally, the comparison of the objective functions (and the terms related to the profit and costs) between six analyzed systems suggests the overall optimal process solution. Beside that, the comparison of reaction rates and permeation rates profiles between several optimal process solutions, provide more insightful information and grounds for the selection of the optimal solution. In conclusion, the results demonstrate significant difference in economic and technological performance. This example demonstrates the new methodology well, and provides the basis for the further development of the suggested concept.

(54a) Your Startup Is a Success -- Now What???

Mike Skowronek, Strata Business Consultants, Houston, TX

YOUR STARTUP IS A SUCCESS -- NOW WHAT??

Mike Skwronek

There are a few universal truths in business life. This is one: You will leave your business someday – you just don't know when. There is only one way an owner can do so successfully. You must create and exit plan as early as possible and stick to that plan as long as you remain in business. Mike Skowronek is a business consultant, Accredited Valuation Analyst (AVA) and Certified Exit Planner (CExP™) who designs exit plans for business owners. The planning process includes coordination with the business owner's other advisors including attorneys, CPAs, investment advisors and insurance professionals to insure that all critical issues are addressed. He will review a proven Seven Step Process™ that combines several elements into a comprehensive written plan including:

- Setting Exit Objectives
- Business and Personal Financial Resources
- Maximizing and Protecting Business Value
- Ownership Transfers to Third Parties
- Ownership Transfers to Insiders
- Business Continuation
- Personal Wealth and Estate Planning

His presentation is designed to make business owners aware of the need for effective planning and the opportunities for a successful exit that proper planning can create.

(54b) The Secret of Success Is There's No Secret

Elizabeth Drake, Compliance Strategies and Solutions, Inc, Houston, TX

Drake's method of success is simple. "Realize that life is short and that you've got one shot," she said. Drake was the only woman graduate with a bachelor's degree in chemistry from East Carolina University in 1984. "The only girl, can you believe that?" She interned in the quality assurance lab of the second largest pharmaceutical company in the world. They offered her the opportunity to continue fulltime, but such a job didn't ignite anything in the passionate lady.

From helping facilities develop green energy, assisting in energy recovery, developing best management practices, gas cleaning and purification, recycling, to finding cost-saving solutions, CS2 has experts with knowledge that literally spans the alphabet—from air pollution control engineering to waste-to-energy operations.

She quit her plant job in 1997. It was a home-based business for a while. She started making cold calls that weren't cold at all and quickly landed some huge jobs. Within six months, she moved into an office building. Her college roommate, Jamie Phillips '87—"she did my taxes"—became her accountant.

Her husband became one of her engineers, and, because of her one-foot-in-front-of another business plan, CS2 was founded with no borrowed money.

CS2's client list is impressive: NASA, Clean Harbors, hazardous waste incinerators, petrochemical companies, pharmaceutical companies, chemical companies, a nuclear facility, and more. And when the petroleum company BP sent out a call to all the best companies to help run its multilayered environmental audits, CS2 was picked for seven of the jobs, outperforming some of the biggest companies in the nation.

"There are no guarantees for anybody. My children don't think that there is anything they can't do. I don't have to tell them." Instead Drake teaches by showing. "You've got to be a giver. You've got to serve others. You've got to make a difference. You've got to be a risk-taker. Take a look at yourself and think, 'I can do this.'"

(54c) What Price to Pay? or What Price to Sell?

Clinton Bogart, Industrial Valuation, C B Richard Ellis, Houston, TX

For an entrepreneur there will be many times that the valuation of their business will be fundamental to the success (or unfortunately the demise) of the business.

- How much to pay to purchase an existing business
- How much to pay to purchase an add on to expand your business
- The bank/lender needs to value your business so they can give you funding
- You have decided to sell your business – what price to offer.
- There is a buyer for the business and are you receiving the right value for your asset.
- Valuation of the ongoing business for ad valorem and other taxes

Understanding what an appraiser is looking at and how they are going to develop the valuation of your plant/facility will help you and the appraiser to get to the "right" value. This will take much of the uncertainly and anxiety out of the appraisal process.

Looking at the value of the assets ("cost approach"), then developing the present value of the future of your business ("income approach") then finally developing comparable sales of similar businesses in the market place will be done by the owner and the appraiser.

Clint Bogart has 20 years of appraising chemical plants, refineries, oil terminals and many other process related assets. This presentation gives business owners and prospective business owners a unique opportunity of find out about the valuation process and how to maximize their valuation.

Avoiding Common Mistakes in Licensing Agreements and Other Technology Agreements

By: Anthony P. Venturino, Esq. , Partner of Novak, Druce + Quigg, LLP Washington, DC

AIChE Spring Meeting - March 15, 2011 Chicago, IL

The views presented here are those of the author and not attributable to his firm or its clients.

I. Technology agreements are more than just licenses

You may encounter many kinds of agreements having technology or other forms of intellectual property as a critical component in the course of starting up. A few such agreements are listed as follows:

- Confidentiality Agreements
- Vendor Supplier Agreements
- Joint Development/Commercialization Agreements
- Exclusive Licenses / Non-exclusive Licenses
- Co-promotion Agreements
- Employee Agreements
- Settlement Agreements
- Joint Venture Agreements
- Contract R&D Agreements
- Consulting Agreements

II. Where Things Go Wrong

Things can go wrong before, during and after the agreement. Here are a few examples of where things can go wrong:

- Evaluation of the technology
- Selection of partners
- Conduct of the negotiations
- The agreement itself
- Implementing and servicing the agreement

Here are classic Licensee mistakes:

- Royalties paid when licensed patent had expired
- Payments made after royalty terms had been satisfied
- Payments made under a trademark license when the trademark is not used
- Patents covered in a license no longer cover products that are currently made, yet royalty is paid against the patent coverage
- Royalties paid at the wrong rate, i.e. one higher or lower than provided Source: *Licensing – A Strategy for Profits*, Edward P. White (1990)

Licensors are no stranger to mistakes. Here are a few classic Licensor mistakes:

- Failure to take action if royalty payments cease when they should not, e.g., prior to expiration of the patents.
- Failure to monitor royalty payments, e.g. sudden reductions or failure to meet minimum royalties that could trigger a change from exclusive to nonexclusive or termination of the licenses.
- Failure to stay in technical contact with the licensee, with the result of licensee not interested in paying royalties, in maintaining quality, or keeping good records.

Licensing – A Strategy for Profits, Edward P. White (1990)

Before you do the deal.

- Know your business goals, strengths and weaknesses
- Know the available protection for the technology
- Have a realistic idea of the value of the technology
- Be open to creative technology agreements
- Admit when you are in over your head and get help
- Do not let paranoia get in the way of a good deal
- Watch out for foreseeable legal and technical problems
- Admit when you are in over your head and get a lawyer or other licensing professional

III. Avoiding Common Traps

A. Define how to handle joint invention of improvements

By US statute joint inventions are jointly owned. It is better to have one party own the invention and, if necessary, grant the other party an exclusive field of use license. Thus, a clause such as the following should be included if a jointly developed invention may occur during performance of the agreement:

> In the event of a joint development my company owns the invention and controls patent prosecution and your company gets an exclusive use license in the field of
> ______________.

B. Keep good records of what the agreement covers or when it expires

The company, it legal department or a law firm should docket the agreement in a database. The docketing record should include: the type of agreement, the parties, the technology covered by the agreement, the agreement deadlines, the expiration date, the contact person and backup contact person, and have a miscellaneous portion for anything unusual.

Preferably an institutional person, for example, a docket clerk or contracts administrator, should be responsible for tracking notices in the docketing system and then relay the information to the business people. The more automatic you can make the timely conveying of such information, the better.

C. Do not let an agreement to agree inadvertently become an agreement

The first time you negotiate with a party, especially if the other party sees this agreement as a one shot deal, be careful not to make records during the negotiations which unexpectedly bind you. Do not accidentally create an agreement.

Summaries of the negotiations signed by both parties can lead to binding agreements even when a clause that might be important to you is missing. Let's take a hypothetical scenario where two parties bring their business people and lawyers to a meeting to negotiate a potential IP license. Mid-afternoon the attorney for the IP owner (potential licensor) had to leave to catch a plane. The potential licensor's business person (let's call him "Bill") had a later plane. The potential licensee kindly offered to take Bill to the airport later so he would not have to waste time in a boring airport. While Bill was by himself with the potential licensee's attorney and business people one of the business people suggested they draft and sign a summary of the day's discussions. They did this and then Bill caught his plane. The next day Bill's company was asked to sign an Assignment document to reflect the assignment they agreed to yesterday. Bill's response was "What assignment?" After a lawsuit the parties settled. The moral of the story is train the business people not to sign anything.

A deal can also inadvertently arise if the parties act like there is an agreement can create an agreement. This sometimes arises where the parties are in a rush to do joint development. While the deal is only a draft they start working together. Sometimes the unsigned draft becomes the *de facto* agreement.

The opposite can also happen. Sometimes you think you have an agreement but in reality you do not, or at least not all the clauses apply. Take the case of *DataTreasury v. Wells Fargo,* 522 F.3d 1368 (Fed. Cir. 2008). In 2003 WMR e-Pin licensed a patent to Wells Fargo Services Corp. a subsidiary of Wells Fargo & Co. The license included an arbitration clause. In 2006 WMR e-Pin assigned 4 patents to DataTreasury. At least one of the 4 patents was a patent of the license. Afterwards DataTreasury sued Wells Fargo & Co. for patent infringement in court. Wells Fargo & Co. moved to dismiss or stay pending arbitration. Under 5th Circuit law, as applied by the Federal Circuit, "to be enforceable, an arbitration clause must be in writing and signed by the party invoking it. *DataTreasury* at 1373, quoting *Westmorland v. Sadoux*, 299 F.3d 462, 465 (5th Cir. 2002). Neither Wells Fargo & Co. nor DataTreasury signed the original patent license. Thus, the court held these parties were not bound by the arbitration clause.

D. Remember important clauses

Clause checklists are important tools when reviewing a license to see if all the necessary clauses are present. However, use the right checklist for the type of agreement and the type of IP. Do not assume the clauses in your standard patent license belong in your new trademark license and *vice versa*. Unlike patents which protect the patent holder, trademarks protect the public by assuring the public a good labeled with a particular trademark will have the characteristics they expect. For example, a PIRELLI brand tire has certain handling properties. Trademark licenses without quality safeguards are termed naked licenses and invalidate the trademark rights.

A short checklist is provided in TABLE 1:

TABLE 1
Introductory Clauses
Definitions
Grant of Rights
Consideration/Compensation
Diligence
Product Quality Control
Advertising
3rd Party Infringement Suits
3rd Party Product Liability Suit
Licensor/ Licensee Suits
Termination
Continuing Obligations

A detailed set of clause checklists for a licensing agreement is available upon request from the author.

E. Keep control of the technology after making the deal

Proprietary technology is a competitive advantage. If the licensor loses control of the technology he loses control of the competitive advantage. Thus, the licensor should determine the level of control he is willing to lose. Maybe the licensor thinks he can permit the licensee to sublicense because he trusts the judgment of licensee's president. What if the president leaves? Thus, the agreement preferably precludes sublicensing or assigning to competitors.

Another common way to lose control is for the licensee to be bought by the licensor's competitor. The Agreement should specify it may be assigned by ONE of the following:

1. Anyone;
2. Goes with the business;
3. Goes with the company;
4. Licensor maintains right of first refusal;
5. Licensor maintains right of first refusal which will not be unreasonably withheld;
6. No one.

F. Promote good communication after making the deal

Most problems arise from misunderstandings, lack of thoroughness, carelessness, or failure to administer the agreement at all, i.e., no one in authority is responsible for administration (LES 1983-84 study).

The lawyer can structure the agreement to include mechanisms for communication between the licensor and licensee so little problems and misunderstandings do not become big ones. You cannot force the business people to communicate but the mechanisms provide fertile soil for communication. For example, the agreement should include a contact person and back-up contact person. The agreement can also mandate times for periodic meetings.

G. Include Clean Termination Provisions

Like romantic relationships, agreement relationships can end badly. Like a newlywed couple, the enthusiastic parties to a new business relationship think nothing will ever go wrong. As a result they do not contemplate the relationship failing. Thus, you need an agreement that acts like a pre-nuptial agreement. Include termination provisions in the agreement to give each party a reasonable way to terminate the agreement. This includes planning for unexpected events, for example, bankruptcy or purchase of the other party. This also includes providing ways to terminate or renegotiate if the economics of the deal no longer make the current agreement viable.

IV. Organize A Licensing Department – A final word

Some companies only occasionally do a license or may be at the beginning of setting up a license project group. In these cases, the company might ask his or her lawyer for advice on how to organize the licensing project group.

The license project group will need the support of a top executive. Without this the license project group will generally not get the resources it needs.

Recognize the different functions to be performed. Then, divide or share these functions depending on available resources, business people, consultants or lawyers:

- Profit Center Group
    - optimizes the licensing in and licensing out to make money for the company
- Support Group
    - coordinates technical, legal and accounting aspects of tech transfer
- Administrative Group
    - handles accounting, record keeping and day to day details of maintaining the license.

V. Conclusion

Agreements are part of managing intellectual property. If you know your goals and avoid the traps your agreements can turn technology into profit.

Production of Dimethyl Carbonate via Reactive Distillation Process

Emmanuel A. Dada[1*], Chandrakant B. Panchal[2], and Richard D. Doctor[2] (1) ChemProcess Technologies (CPT), LLC, League City, TX 77573, (2) Energy System Division, Argonne National Laboratory, Argonne, IL 60439.
*: Presenter and corresponding author, emmanuel.dada@ymail.com

Abstract:

Dimethyl Carbonate (DMC) is an environmentally benign and biodegradable product. Because of its high oxygen content, DMC is an idea additive for gasoline to replace methyl tert-butyl ether (MTBE). DMC has many applications in the synthesis of polycarbonates, production of methyl isocyanate and as a replacement of halogenated solvents. It is extensively used as a methylating and carbonylating agent. Dimethyl Carbonate has recently becomes an important solvent in the lithium ion battery solutions and other electrochemical and coating sectors.

The early commercial production of DMC via phosgenation of methanol has many draw backs especially the toxicity of phosgene and is being phased out and replaced by other methods. Among these new routes of producing DMC are the oxidative carbonylation of methanol involving CO, transesterification using propylene carbonate or ethylene carbonate to co-produce DMC and propylene glycol or ethylene glycol, and methanolysis of urea. The merits of each of the new routes of producing DMC will be presented especially the systematic choice of entrainers in the extractive distillation step to separate the DMC-Methanol azeotrope common to all the three routes.

Introduction:

DMC has received significant research efforts in recent years due to its versatile chemical properties and commercial applications. It was first manufactured in the 1910s from methanol and phosgene. This process has many draw backs especially the toxicity of phosgene and is being phased out and replaced by other methods. One of the most important non-phosgene routes which is of the focus for our project is the transesterification of ethylene carbonate with methanol to produce dimethyl carbonate (DMC) carried out in a reactive distillation column. The common separation issue to most of the routes is the separation of the azeotrope mixture of methanol and dimethyl carbonate to make high purity DMC product. This issue is being addressed in our project with a focus to use a heat-integrated reactive distillation method couple with a heat-integrated extractive distillation for high purity DMC product that will be an economically promising and energy savings route.

Design of the Extractive Distillation:

The choice of the best entrainers that is environmentally acceptable, safe, and is relatively cheap is very critical for this process. In this respect we have considered the residue curve maps and the influence of some entrainers on the activity coefficients (related to the separation factor-relative volatility) of the Methanol-DMC azeotrope mixture as shown below figures and table 1. The maps were generated using ICAS software version 13 (2010). Details of the entrainers' selection process will be presented.

Table1: Relative Volatility of Methanol to Dimethyl Carbonate at Infinite Dilution

	Azeotrope Feed plus 50% Aniline, Fig 4	Azeotrope Feed plus 50% Phenol, Fig 1	Azeotrope Feed plus 80% Aniline, Fig 5	Azeotrope Feed plus 80% Phenol, Fig 2	0% Aniline, 0% Phenol, Figure 3
Activity Coeffcient of Methanol at infinite dilution, GammaI	2.3	1.91	2.1	1.1	2.43
Activity coeffcient of DMC at infinite dilution, GammaJ	0.93	0.72	0.81	0.3	0.99
Selectivity of I,J =GammaI/GammaJ	2.47	2.65	2.59	3.67	2.45
Relative Volatility = Selectivity of I,J * Ratio of Pure Comp. Vap Pressure (I to J)	6.35	6.81	6.65	9.41	6.28

Table 2: List of Potential Entrainers for Dimethyl Carbonate-Methanol Mixture

Name	Normal Boiling Point, K	Normal Melting Point K	Dipole Moment Debye	Solubility Parameter (MPa)^0.5	Reference
Aniline C_6H_7N	457.15	267.13	1.5288	24.1171	Hsu, K. et al (2010)
Phenol C_6H_6O	454.99	314.06	1.4508	24.6326	Wang, S.J. et al (2010)
Ethylene Glycol $C_2H_6O_2$	470.45	260.15	2.3082	33.7033	Hsu, K. et al (2010)
Ethylene Carbonate $C_3H_4O_3$	511.15	309.55	4.5095	30.0352	Himmele, W. et al (1979)
Diethyl Oxalate $C_6H_{10}O_4$	458.85	232.55	2.4897	20.1533	Ryu and Gelbein (2002)
Chlorobenzene C_6H_5Cl	404.87	227.95	1.6906	19.3453	Li and Zhang (2006)
Benzonitrile C_7H_5N	464.25	260.45	4.1796	21.7958	Himmele, W. et al (1979)
N,N-dimethylaniline $C_8H_{11}N$	467.25	275.55	1.6798	20.0755	Himmele, W. et al (1979)
Water	373.15	273.15	1.8495	47.8127	Ginnasi , A. et al (1976)

Source of thermophysical properties is ICAS Software version 13 (2010)

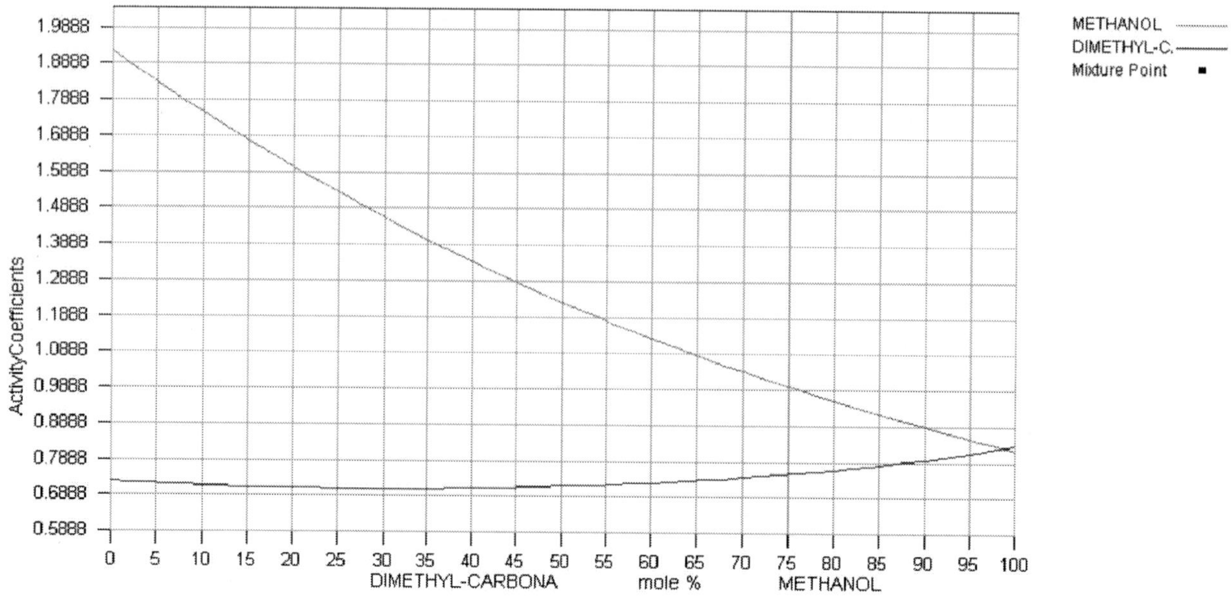

Figure 1: Activity coeff. at 1 atm , 370K , 50% mol phenol with azeotrope feed 87.06 mol MeOH;12.94 mol DMC

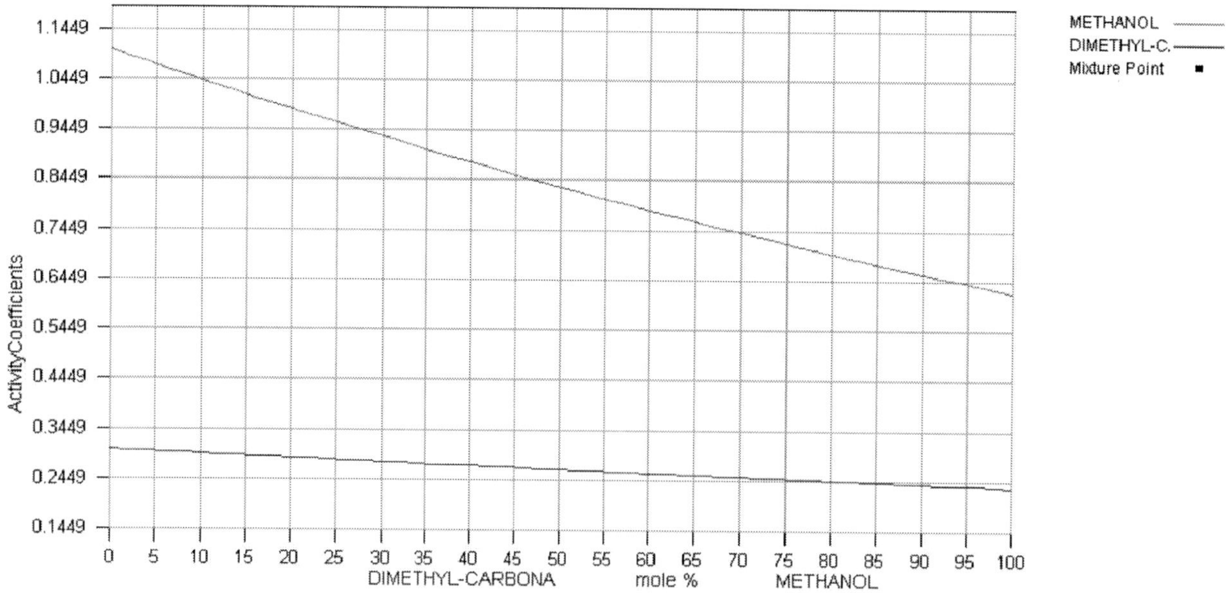

Figure 2: Activity coeff. at 1 atm , 370K , 80% mol phenol with azeotrope feed 87.06 mol MeOH;12.94 mol DMC

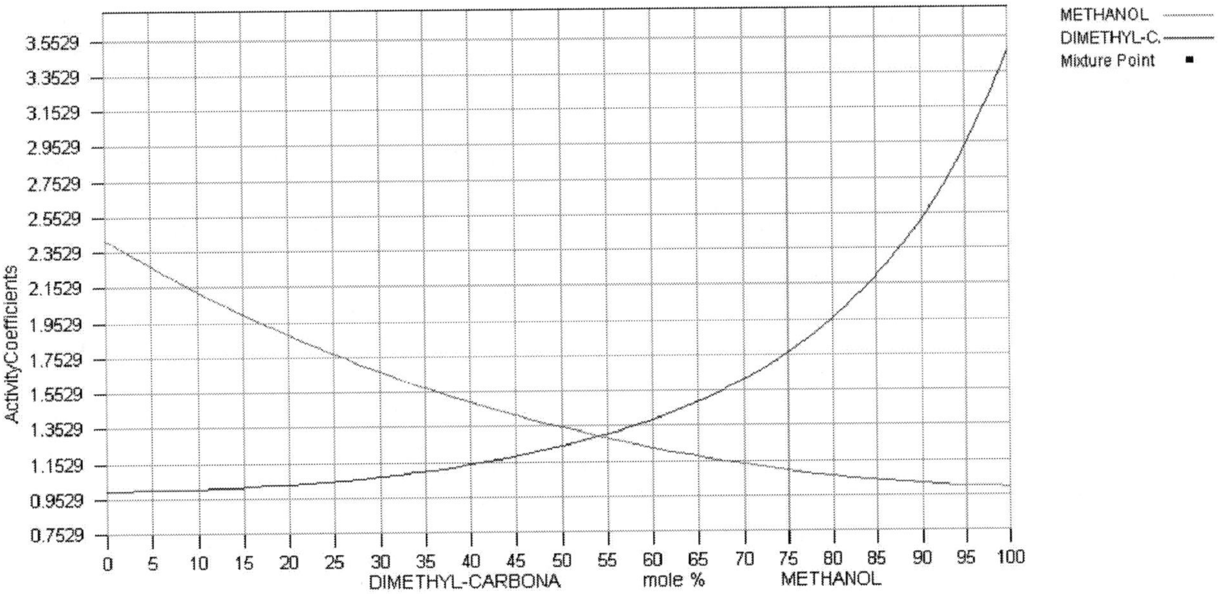

Figure 3: Activity coeff. at 1 atm , 370K , 0 % mol aniline with azeotrope feed 87.06 mol MeOH;12.94 mol DMC

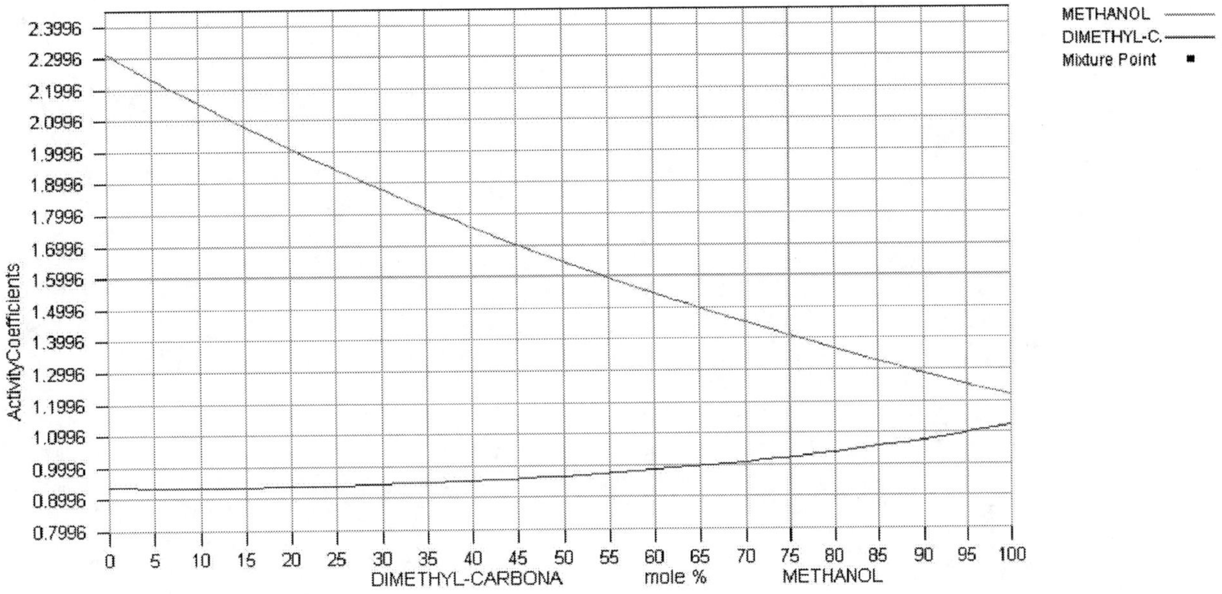

Figure 4: Activity coeff. at 1 atm , 370K , 50% mol aniline with azeotrope feed 87.06 mol MeOH;12.94 mol DMC

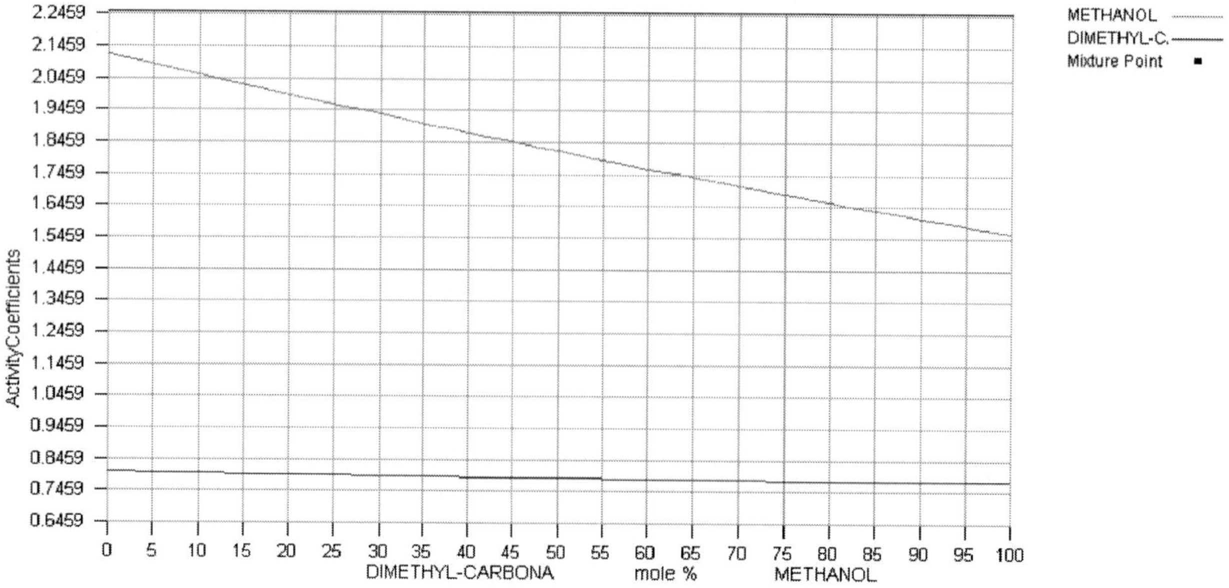

Figure 5: Activity coeff. at 1 atm , 370K , 80% mol aniline with azeotrope feed 87.06 mol MeOH;12.94 mol DMC

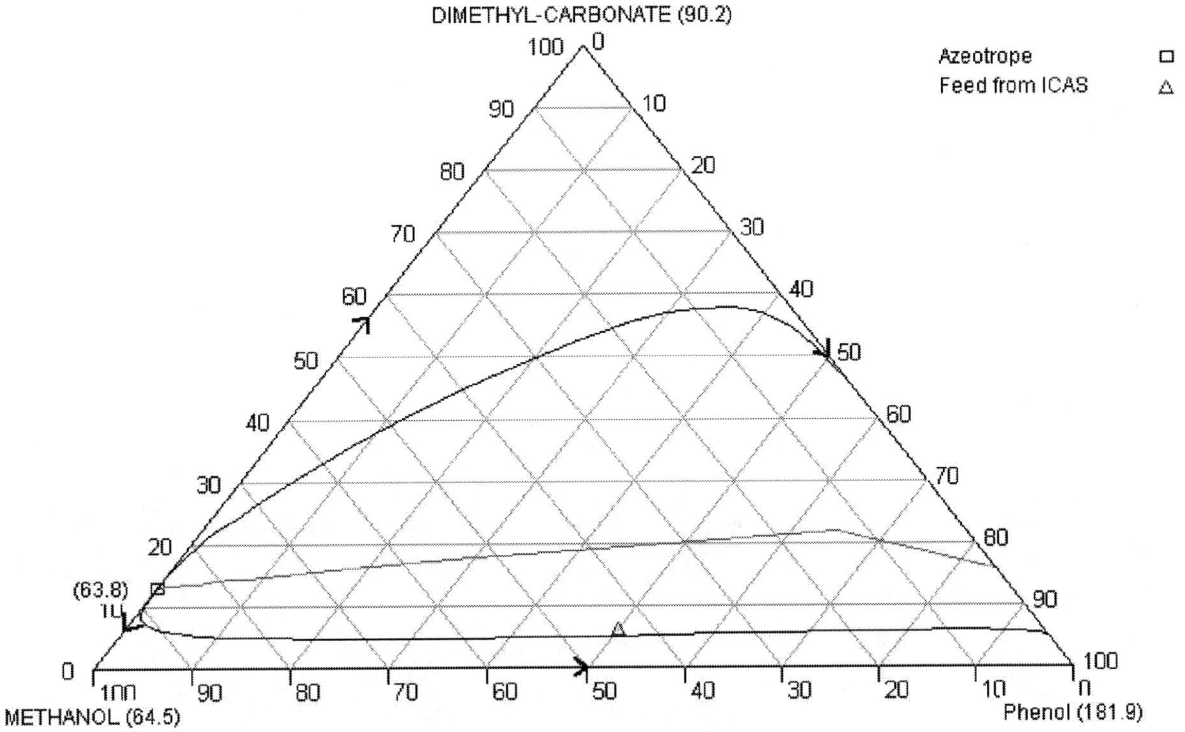

______________ Upper Binodal Curve at 1= 30%, 2=50%, and 3= 20%

______________ Lower Binodal Curve at 1=50%, 2=5%, 3= 45%

______________ Distillation Boundary

Feed from ICAS: 1=87.06 mole, 2= 12.94 mole , and 3 = 100 mole (Azeotrope feed MeOH-DMC)

1=mol % Methanol, 2 = mole % DMC , 3 = mol % Phenol

Figure 6: Residue Curve Map for Extractive Distillation of Methanol-DMC-Phenol System at 1 atm

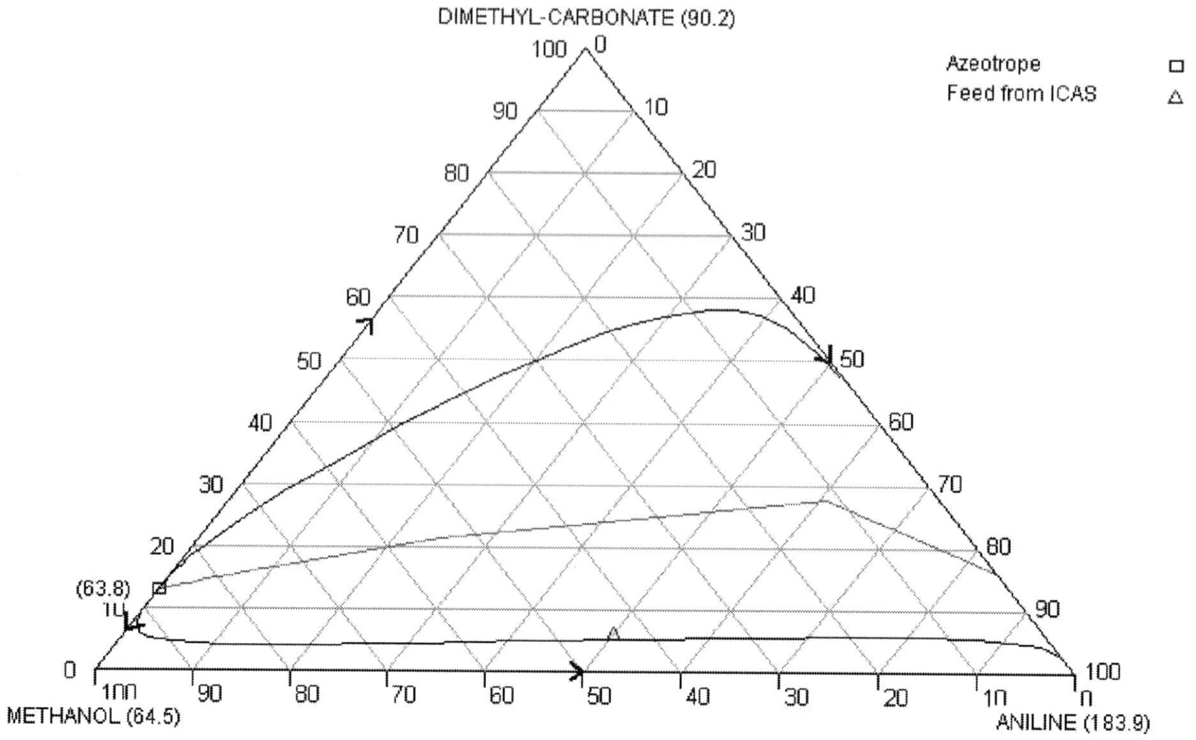

_______________ Upper Binodal Curve at 1= 30%, 2=50%, and 3= 20%

_______________ Lower Binodal Curve at 1=50%, 2=5%, 3= 45%

_______________Distillation Boundary

Feed from ICAS: 1=87.06 mole, 2= 12.94 mole , and 3 = 100 mole (Azeotrope feed MeOH-DMC)

1=mol % Methanol, 2 = mole % DMC , 3 = mol % Aniline

Figure 7: Residue Curve Map for Extractive Distillation of Methanol-DMC-Aniline System at 1 atm

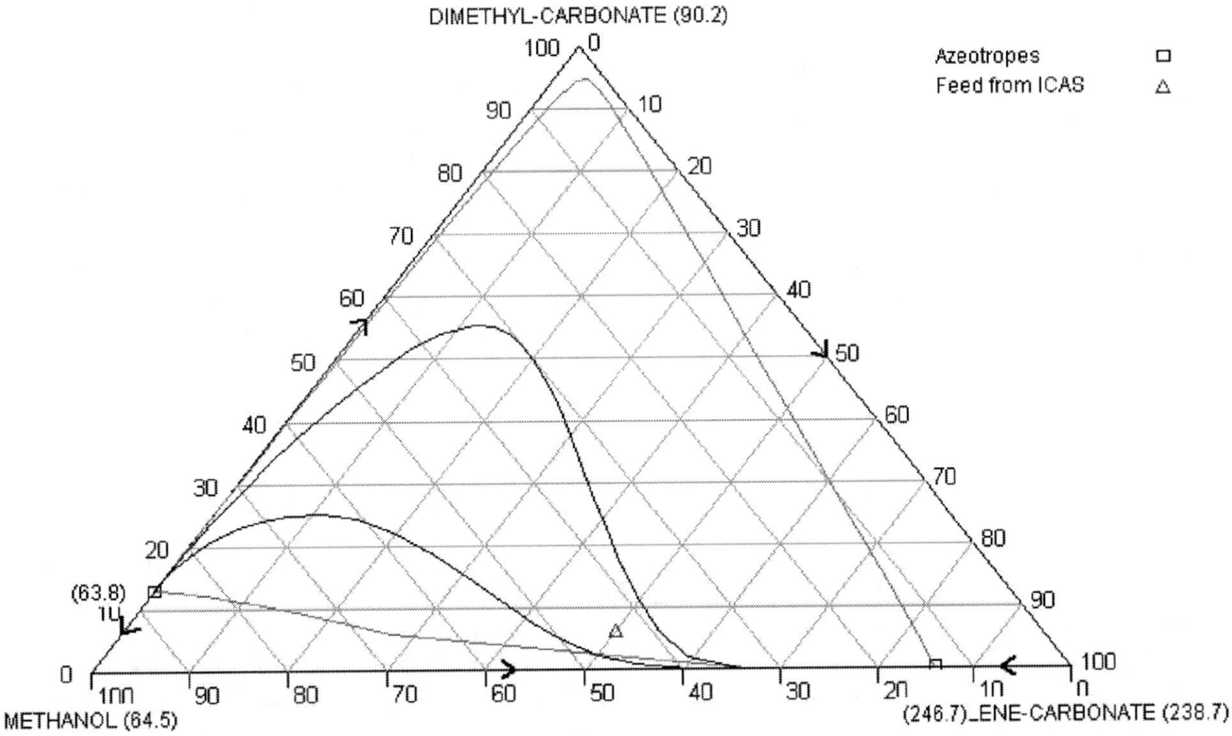

Upper Binodal Curve at 1= 30%, 2=50%, and 3= 20%

Lower Binodal Curve at 1=50%, 2=5%, 3= 45%

Distillation Boundary

Feed from ICAS: 1=87.06 mole, 2= 12.94 mole , and 3 = 100 mole (Azeotrope feed MeOH-DMC)

1=mol % Methanol, 2 = mole % DMC , 3 = mol % Ethylene Carbonate

Figure 8: Residue Curve Map for Extractive Distillation of Methanol-DMC-Ethylene Carbonate System at 1 atm.

Conclusion:

The focus of this project is the application of a heat-integrated reactive distillation couple with a heat-integrated extractive distillation process for the production of a high purity dimethyl carbonate by transesterification of ethylene carbonate with methanol to achieve energy reduction/savings and is economically promising.

References:

Fang, Y. J.; Qian, J. M. Isobaric vapor-liquid equlibria of binary mixtures containing the carbonate group -OCOO-. *J. Chem. Eng. Data* **2005**, *50*, 340–343.

Fang, Y.; Xiao, W.; Experimental and modeling studies on a homogeneous reactive distillation system for dimethyl carbonate synthesis by transesterification. Separation and Purification Technology 34 (2004) 255-263
Fang, Y. J.; Xiao, W. D. Experimental and modeling studies on a homogeneous reactive distillation system for dimethyl carbonate synthesis by transesterification. *Sep. Purif. Technol.* **2004**, *34*, 255–263.

Foucher, E. R.; Doherty, M. F.; Malone, M. F. Automatic screening of entrainers for homogeneous azeotropic distillation. *Ind. Eng. Chem. Res.* **1991**, *30*, 760–772.

Gmehling, J.; Mollmann, C. Synthesis of distillation processes using thermodynamic models and the Dortmund data bank. *Ind. Eng. Chem. Res.* **1998**, *37*, 3112–3123.

Ginnasi, A., Passoni, G.; Extractive distillation of a dimethyl carbonate feed with water. Patent US 3,963,586 (196)

Harmsen, G. Jan, Reactive Distillation: The front-runner of industrial process intensification. A full review of commercial applications, research, scale-up, design and operation. Chemical Engineering and Processing 46(2007) 774-780

Himmele, W. et al; Preparation of pure dimethyl carbonate, patent US 4,162,200 (1979)

Hiwale, R.S.; Bhate, N. V.; Mahajan, Y.S.; Mahajani, S. M.; Industrial applications of reactive distillation:Recent trends. International Journal of Chemmical Reactor Engineering, Vol 2(2004), Review R1 pp 1- 52

Hsu, K.; Hsiao, Y.; Chien, I.; Design and control of dimethyl carbonate-methanol separation via extractive distillation in dimethyl carbonate reactive-distillation process. Ind. Eng. Chem. Res. 210, 49, 735-749

http://www.chemsystems.com/reports/search/docs/abstracts/99S6-abs.pdf ; Developments in Dimethyl Carbonate Production Technologies PERP-99-S6 (May. 2000)

Hu, Wang-Hu; Shen, L; Zhao, L, Measurement of vapor-liquid equilibrium of binary mixtures of phenol-dimethyl carbonate and phenol-methanol at 101.3 kPa. Fluid Phase Equilibria 219 (2004) 265-268

ICAS Software version 13, CAPEC, Technical University of Denmark (2010)

Kossack, S.; Kramer, K.; Gani, R.; Marquardt, W. A systematic synthesis framework for extractive distillation processes. *Chem. Eng. Res. Des.* **2008**, *86*, 781–792.

Lek-utaiwan, P; Suphanit, B; Mongkolsiri, N.; Gani, R.; Integrated design of solvent-based extractive separation processes including experimental validation. 19th European Symposium on Computer Aided Process Engineering-ESCCAPE 19 J.Jezowski and J. Thullie (Editors), 2009 Elsevier

Li, C.; Zhang, S. Environmental Benign design of DMC production process. Trans IChemE, Part A, Chemical Engineering Research and Design, 2006, 84(A1):1-8

Nishihira, K., Yoshida, S; Tanaka, S., Continuous process for preparing dimethyl carbonate. Patent US 5,124,185 (1993)

Pasternak, M., et al; Membrane separation of organic liquids, patent US 4,960,519 (1990).

Rodriguez, A.; Canosa, J.; Dominguez, A.; Tojo, J. Vapour-liquid equilibria of dimethyl carbonate with linear alcohols and estimation of interaction parameters for the UNIFAC and ASOG method. *Fluid Phase Equilib.* **2002**, *201*, 187–201.

Rodriguez-Donis, I.; Gerbaud, V.; Joulia, X. Entrainer selection rules for the separation of azeotropic and close-boiling-temperature mixtures by homogeneous batch distillation process. *Ind. Eng. Chem. Res.* **2001**, *40*, 2729–2741.

Ryu, J. V.; Gelbein, A. P.; Process and catalyst for making dialkyl carbonates, patent US 6,392,078 B1 (2002)

Sachs, H. M.; Preparation of alkylene carbonates. Patent US 4,786,741 (1988)

Wang, S. J.; Yu, C. C.; Huang, H. P. Plant-wide design and control of DMC synthesis process via reactive distillation and thermally coupled extractive distillation *Comput. Chem. Eng.* 34(2010) 361-373

Wang, X; Yang, B.; Wang, D.; Zhai X.; Reactive rectifying for producing dimethyl carbonate. Chemical Engineering Journal 122 (2006) 15-20

(77b) Reactive Distillation Process for Glycerol Ethers Production

Elena Vlad[1], **Costin Sorin Bildea**[2], Marina Mihalachi[1] and Grigore Bozga[1], (1)Department of Chemical Engineering, University Politehnica of Bucharest, Bucharest, Romania, (2)Department of Chemical Engineering, University POLITEHNICA Bucharest, Bucharest, Romania

Glycerol is produced as by-product of biodiesel production in amounts that are equivalent to approximately 10 % wt. of the total product. As a result to the increasing availability, the market price of glycerol has dropped rapidly and new uses for glycerol need to be found. Di- and tri-ethers of glycerol are compounds soluble in diesel and biodiesel, improving the quality of the fuel. They diminish the emissions of particulate matter, carbon oxide and carbonyl compounds and decrease the cloud point of diesel fuel when combined with biodiesel. Therefore, ethers of glycerol are interesting alternatives to commercial oxygenate additives.

In this contribution, we demonstrate the feasibility of a reactive distillation process for production of glycerol ethers by etherification of glycerol with iso-butene catalyzed by heterogeneous acid catalysts, such as Amberlist 35. The process is performed in a reactive distillation column, with few un-reactive trays placed above and below the reactive zone. Glycerol is fed as liquid at the top of the reactive zone. Iso-butene enters in the column as a vapor stream at the bottom. The bottom product contains the di- and tri-ethers of glycerol which are the main products. The vapor outlet stream contains the mono-ether and the excess of iso-butene, which are separated in a decanter. The mono-ether is returned as reflux, while the excess of iso-butene is recycled. The detailed design is performed using AspenPlus. The stability and robustness in operation is checked using rigorous dynamic simulation in AspenDynamics.

(77c) Advanced Reactive Distillation Concepts for the Indirect Hydration of Cyclohexene to Cyclohexanol

Aspi K. Kolah[1], Lars Peereboom[1], Carl T. Lira[1], Jing Huang[2], C.B. Panchal[2], Robert W. Lyczkowski[2], Emmanuel A. Dada[3], Richard D. Doctor[2] and Dennis J. Miller[1], (1)Chemical Engineering and Materials Science, Michigan State University, East Lansing, MI, (2)Energy System Division, Argonne National Laboratory, Argonne, IL, (3)ChemProcess Technologies (CPT), LLC, League City, TX

Cyclohexanol is a very important bulk chemical feedstock for the polymer industry, primarily as a precursor to nylon but also a plasticizer component and solvent. The conventional process for cyclohexanol synthesis is either cyclohexane oxidation, hydrogenation of phenol, or the direct hydration of cyclohexene. The direct hydration route, which is the simplest in principle, suffers from limited solubilities of the cyclohexene-water pair, thus requiring large amounts of catalyst and high residence times.

An alternative to the direct hydration of cyclohexene is the indirect hydration process where acetic acid is used as the reactive entrainer. In the first step of this process cyclohexene from the inlet mixture of cyclohexene and cyclohexane reacts with acetic acid to form cyclohexyl acetate, which is hydrolyzed with water to cyclohexanol in the second step.

In this paper we present the reaction of cyclohexene from a mixture of cyclohexene and cyclohexane with acetic acid to form cyclohexyl acetate in the MSU pilot scale reactive distillation column packed with Katapak-SP11 structured packing. Effect of flow rates, reflux ratio and column geometry have been experimentally investigated. Column performance is predicted using RADFRAC in Aspen Plus process simulation software with appropriate kinetic parameters derived from experimental batch kinetic studies. Additionally, single- and two-phase flow through a Katapak-SP11 structured packing was modeled using ANSYS FLUENT 12.1. The purpose of this analysis is to determine the liquid flow distributions interior and exterior to the catalyst baskets, catalyst wetting, and hence reaction effectiveness. Advanced concepts such as heat integration are incorporated into experimental and simulation results. Plant scale simulations are presented.

A review of techniques for the process integration of coupling exothermic and endothermic reactions

Mohammad Reza Rahimpour;
Davood Iranshahi;
majid raoof Dehnavy;
Fatemeh Allahgholipour;

Presentation ID:212277

School of Chemical and Petroleum Engineering, Shiraz University, P.O. Box 71345, Shiraz, Iran

Keywords: Couple, Endothermic, Exothermic, Process integration

Corresponding author. Tel.: +98 711 2303071; fax: +98 711 6287294.
E-mail address: rahimpor@shirazu.ac.ir (M.R. Rahimpour).

1. Introduction

Reducing environmental pollutions, reducing capital and operating costs of processes and increasing efficiency and outputs of processes are significant terms which should be noticed in this new world. Process integration (PI), which is one of the most significant methodology in chemical engineering and process technology, can prove these aims by combining several processes to reduce consumption of the energy and material or environmental emissions [1] and also it can decrease the size of equipment by several orders of magnitude leading to substantial savings in capital cost or improvement of intrinsic safety. The study on process integration started in the late 1970s and early 1980s. [2]

PI includes hierarchical design methods, knowledge-based systems, numerical and graphical techniques and pinch analysis [3]. Pinch analysis designs a process to minimize energy consumption and enhance energy efficiency. In this way multifunctional reactor is a new one [4]. The best field of using multifunctional reactors is the coupling of endothermic and exothermic reactions. In this way, an exothermic reaction(s) produces the source of heat to drive the endothermic reaction(s). The distinction between net source and net sink regions is a key point in the pinch analysis [3]. In this paper, we are investigating review of coupling concept with focus on the reactors, and also, consequences of previous researches in this field are evaluated.

2. Categorize the kinds of coupling of exothermic and endothermic [5]:

2. a- Direct coupling (direct coupled adiabatic reactor)
2. b- Regenerative coupling (reverse-flow reactor)
2. c- Recuperative coupling (counter current heat exchanger reactor and co-current heat exchanger reactor).
2. d- Membrane reactor

2. a) In directly coupled reactors, exothermic and endothermic reactions are in a same bed. There is a direct heat transfer between them, and also, both reactions occur simultaneously. Direct coupling adiabatic reactor (DCAR) can be also classified into simultaneous DCAR (SIMDCAR) and sequential DCAR (SEQDCAR). In SIMDCAR, uniformly mixed endothermic and exothermic catalysts make the bed of it. As it can be seen in Fig.1 (a) cayalyst are in a same bed and feed contains both of reactants (exothermic reactant and endothermic one). Fig.1 (b) illustrates a SEQDCAR reactor.

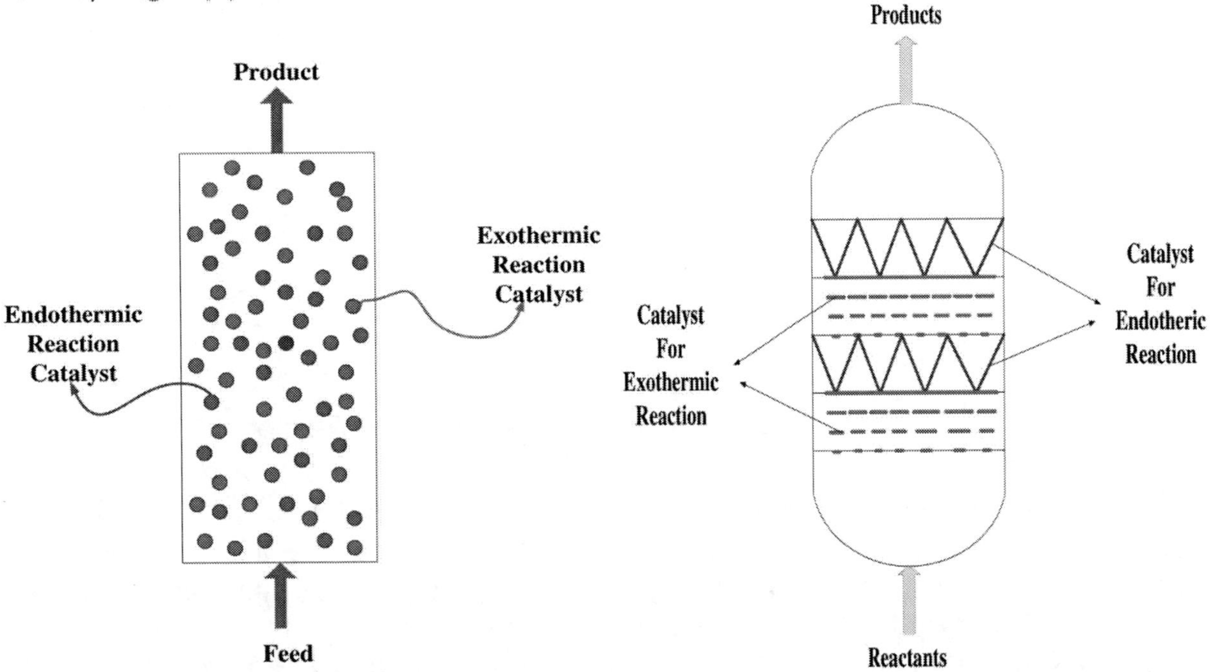

Figure.1 (a) Simultaneous direct coupled adiabatic reactor and (b) sequential direct coupled adiabatic reactor.

2. b) In regenerative coupling reactor, both reactions are in a same place but separated in time. There are two steps and exothermic (or endothermic) and endothermic (or exothermic) reactants enter the reactor respectively. Regenerative coupling reactors are well adaptable for weakly exothermic reactions such as the catalytic purification of exhaustive streams are polluted by volatile organic compounds (Kulkurani, 1996). In the first half of the cycle, the exothermic reaction takes place. Then, the flow is reversed in the next half of the cycle and the endothermic reaction occurs by using of the stored energy in the bed during the previous exothermic half cycle.

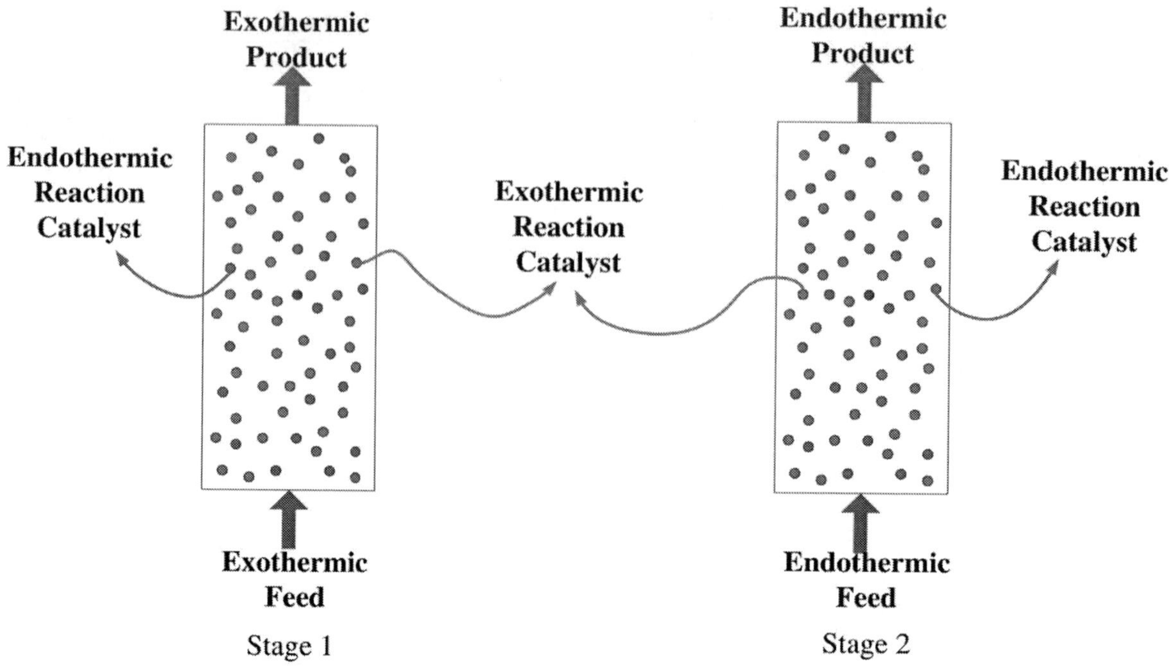

Figure.2 Regenerative coupling reactor.

2. c) In recuperative coupling reactor that is showed in fig.3, there are two separated spaces for two reactions. Tube side is for endothermic reaction and shell side is for exothermic one and heat is transferred indirectly between shell and tube side. In this kind of reactor, the reactors for the endothermic reactions are fed through contiguous passages of the reactor [6, 7]. These passages are either channels in a monolith or catalyst coated. Koga and Watanabe (1991) described a coupling of the catalytic combustion and reforming, which took place in alternate channels. Igurshi et al., (1992) described a wall reactor that supplied heat to endothermic reaction with a heating medium.

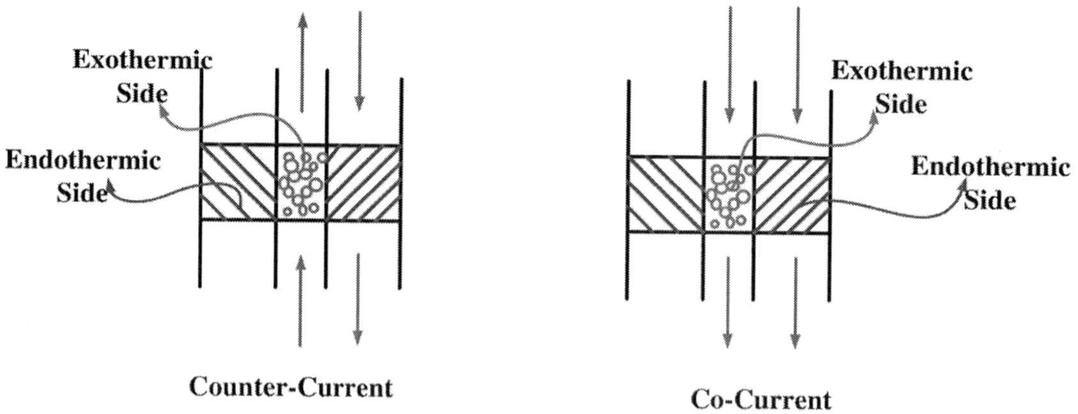

Figure.3 Recuperative coupling reactor.

2. d) In membrane reactor, exothermic reaction carries out in the tube side, and endothermic reaction occurs in through shell side and membrane. A simplified schematic of the membrane reactor is illustrated in Fig.4.

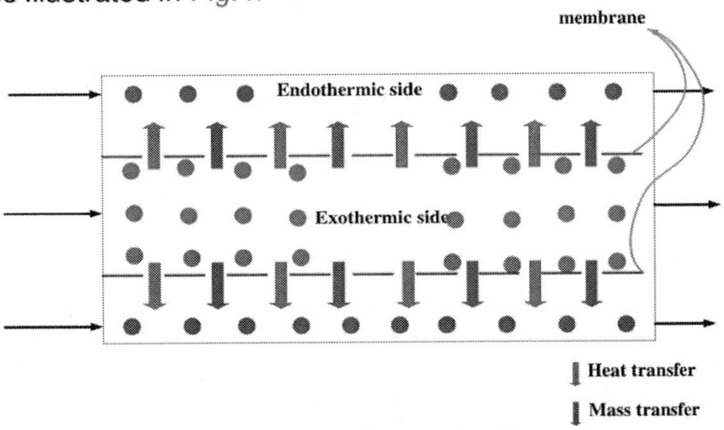

Figure.4 Membrane reactor.

3. Main Table

In this section, we demonstrate examples of the various studies on coupling concept in the Main Table.

No.	Sec.	Reaction	Reaction kinetic	ΔH_{298} (Kj/mol)	T (K)	P	Phase	Catalyst	Author	Ref
1	1-1	Methane combustion reaction	$CH_4 + 2O_2 \rightarrow CO_2 + 2H_2O$	(-)	1000	3 bar	G	Pt / Al_2O_3	Annaland et al.	9, 10, 11, 12
	1-2	Dehydrogenation Propane to propylene	$C_3H_8 \leftrightarrow C_3H_6 + H_2$	130	823-923	1 atm	G			
2	2-1	Methanol Synthesis	$CO + H_2 \rightarrow CH_3OH$	(-)	500	76.98 bar	G	CuO / ZnO / Al_2O_3	Rahimpour et al.	6, 13, 14, 15, 23
	2-2	Dehydrogenation of Cyclohexane to benzene	$C_6H_{12} \rightarrow C_3H_6 + 3H_2$	+206.2.	423-523	101.3 KPa	G	Pt / Al_2O_3		

3	3-1	Dehydrogenation methanol to (DME)	$2CH_3OH \rightarrow CH_3OCH_3 + H_2O$	-23.4	535	18.18 bar	G	γ / Al_2O_3	Rahimpour et al.	16
	3-2	Dehydrogenation of Cyclohexane to benzene	$C_6H_{12} \rightarrow C_3H_6 + 3H_2$	206.2	423-523	101.3 KPa	G	Pt / Al_2O_3		
4	4-1	Hydrogenation of nitrobenzene to aniline	$C_6H_5NO_2 + 3H_2 \rightarrow C_6H_5NH_2 + 2H_2O$	-443	780	1100 KPa	G	Pd / Al_2O_3	Iranshahi et al.	8
	4-2	Naphtha reforming	(CnH2n) ↔ (CnH2n-6) + 3H2	71038.06	777	3703 KPa	G	Pt / Re / Al_2O_3		
			(CnH2n)+H2 ↔ (CnH2n+2)	-36395.33						
			(CnH2n) + n/3H2 → (C1–C5)	-51939.31						
			(CnH2n+2) + (n−3)/3H2 → (C1–C5)	-56597.54						
5	5-1	Hydrogenation of nitrobenzene to aniline	$C_6H_5NO_2 + 3H_2 \rightarrow C_6H_5NH_2 + 2H_2O$	−443	300 - 700	1.1 bar	G	Pd / Al_2O_3	Abo-Ghandar et al. & Qin et al.	17, 18, 19
	5-2	Dehydrogenation of ethylbenzene to styrene	$C_6H_5CH_2 \leftrightarrow C_6H_5CHCH_2 + H_2$	+117.6	300 - 700	4.5 bar	G	Fe_2O_3		

6	**6-1**	The CO2 methanation	$CO_2 + 4H_2 \leftrightarrow CH_4 + 2H_2O$	-165	300 – 700	0.1-1 MPa	G	Ni	Qin et al.	19
	6-2	Dehydrogenation of ethylbenzene to styrene	$C_6H_5CH_2 \leftrightarrow C_6H_5CHCH_2 + H_2$	+117.6	300 - 700	4.5 bar	G	iron oxide		
7	**7-1**	Methane oxidation	$CH_4 + 2O_2 \rightarrow CO_2 + 2H_2O$	-802	300 – 700	15 – 25 bar	G	Ni	Abashar & Elnashei	20, 21, 22
	7-2	Methane steam reforming	$CH_4 + H_2O \rightarrow CO + 3H_2$	+206			G			

The most significant authors related with coupling of exothermic and endothermic reactions field are Annaland, Abo-Ghander, Rahimpour, Abashar, Elnashaie. Work of these head authors were extended or used by other researchers.
By investigation on the authors work which are explained in the Main Table, It can be seen most of them have utilized membrane and recuperative model in their work

4. Conclusion

With this literature review we aimed to give information about coupling the exothermic and endothermic reactions to researchers. The papers that were listed in the Main Table revealed values of the coupling of exothermic and endothermic reactions and coupled reactors. Nowadays, process integration, particularly pinch analysis is the inseparable part in the chemical processes. Therefore, the coupling techniques as a pinch analysis should be taken into consideration and used in chemical industries. As it can be seen in all work related with this field, the short distance between heat sink and heat source could increase efficiency of the heat transfer and incorporating two separated processes with a coupled reactors could reduce size of the operations. And also there are still many reactions which can be coupled with each other. Studying on new coupling of reactions is needed in order to achieve PI's aims. This shows priority of the novel models (coupled reactors configuration) to the conventional configurations.

References

1. F. Friedler (2010), Process integration, modeling and optimization for energy saving and pollution reduction, Applied Thermal Engineering (2008), doi: 10.1016/ j.applthermaleng. 04.030
2. R. Smith(2000), State of the art in process integration, Applied Thermal Engineering (2008), 20 1337-1345
3. A.P. Rossiter(2003), Succeeding in process integration, Process Industries Expo User Conference, March 31.
4. L. Lavric, V. Plesu, J. De Ruyck(2005), Chemical reactors energy integration through virtual heat exchangers-benefits and drawbacks, Applied Thermal Engineering1033-1044.
5. Ramaswamy C. Ramaswamy(2006), Steady state and dynamic reactor models for coupling exothermic and endothermic reactions, doctor's degree at the University of Washington, May,.
6. M.H. Khademi, M.R. Rahimpour, A. Jahanmiri(2010), Differential evolution (DE) strategy for optimization of hydrogen production, cyclohexane dehydrogenation and methanol synthesis in a hydrogen-permselective membrane thermally coupled reactor, Int.J.Hydrogen Energy 35 1936 – 1950
7. G. Kolios, J. Frauhammer, G. Eigenberger(2001), A simplified procedure for the optimal design of autothermal reactors for endothermic high-temperature reactions., Chem. Eng. Sci. 56 351-357
8. D. Iranshahi, A.M. Bahmanpour, E. Pourazadi, M.R. Rahimpour(2010), Mathematical modeling of a multi-stage naphtha reforming process using novel thermally coupled recuperative reactors to enhance aromatic production, Int.J.Hydrogen Energy 35 10984-10993
9. M.van Sint Annaland, J.A.M. Kuipers, W.P.M.van Swaaij56 (2001), Safety analysis of switching between reductive and oxidative conditions in a reaction coupling reverse flow reactor. Chem.Eng.Sci. 1517-1524
10. M.van Sint Annaland, H.A.R. Scholts, J.A.M. Kuipers, W.P.M.van Swaaij57 (2002), A novel reverse flow reactor coupling endothermic and exothermic reactions. Part I: comparison of reactor configurations for irreversible endothermic reactions, Chem.Eng.Sci. 833 – 854
11. M.van Sint Annaland, H.A.R. Scholts, J.A.M. Kuipers, W.P.M.van Swaaij57 (2002), A novel reverse flow reactor coupling endothermic and exothermic reactions. Part II: Sequential reactor configuration for reversible endothermic reactions, Chem.Eng.Sci. 855 – 872
12. M.van Sint Annaland, R.C. Nijssen57 (2002), A novel reverse flow reactor coupling endothermic and exothermic reactions: an experimental study, Chem.Eng.Sci. 4967 – 4985
13. M.H. Khademi, A. Jahanmiri, M.R. Rahimpour34 (2009), A novel configuration for hydrogen production from coupling of methanol and benzene synthesis in a hydrogen-permselective membrane reactor, Int.J.Hydrogen Energy 5091 – 5107
14. M.H. Khademi, P. Setoodeh, M.R. Rahimpour, A. Jahanmiri34 (2009), Optimization of methanol synthesis and cyclohexane dehydrogenation in a thermally coupled reactor using differential evolution (DE) method, Int.J.Hydrogen Energy 6930 – 6944
15. M.R. Rahimpour, E. Pourazadi(2010), A Comparison of hydrogen and methanol production in a thermally coupled membrane reactor for co current and counter current flows, Int.J.Energy Research DOI: 10.1002/er.1744
16. M. Farsi, M.H. Khademi, A. Jahanmiri, M.R. Rahimpour(2010), Novel Recuperative Configuration for Coupling of Methanol Dehydration to Dimethyl Ether with Cyclohexane Dehydrogenation to Benzene, Ind. Eng. Chem. Res. 49, 4633–4643

17. N. S.Abo-Ghander, F. Logist, J.R. Grace, J.F.M. V. Impe , S.S.E.H. Elnashaie, C.J. Lim(2010)
, Optimal design of an autothermal membrane reactor coupling the dehydrogenation of ethylbenzene to styrene with the hydrogenation of nitrobenzene to aniline, Chem.Eng.Sci
18. N.S. Abo-Ghandera, J.R. Gracea, S.S.E.H. Elnashaie, C.J. Lim63 (2008), Modeling of a novel membrane reactor to integrate dehydrogenation of ethylbenzene to styrene with hydrogenation of nitrobenzene to aniline, Chem.Eng.Sci, 1817 – 1826
19. Z. Qin, J. Liu, A. Sun, J. Wang(2003), Reaction Coupling in the New Processes for Producing Styrene from Ethylbenzene, Ind. Eng. Chem. Res. 42, 1329-1333
20. M.E.E. Abashar, S.S.E.H. Elnashaie(2007), Feeding of oxygen along the height of a circulation fast fluidized bed membrane reactor for efficient production of hydrogen, Trans IChemE, Part A, Chemical Engineering Research and Design, 85(A11): 1529–1538
21. Y. Chen, H. Xu, X. Jin, G. Xiong116 (2006), Integration of gasoline prereforming into autothermal reforming for hydrogen production, Catalysis Today 334–340
22. M.E.E. Abashar, F.M. Alhabdana, S.S.E.H. Elnashaie33 (2008), Discrete injection of oxygen enhances hydrogen production in circulating fast fluidized bed membrane reactors, Int.J.Hydrogen Energy 2477 – 2488
23. M.R. Rahimpour, F. Rahmani, M. Bayat, E. Pourazad(2010), Enhancement of simultaneous hydrogen productionand methanol synthesis in thermally coupleddouble-membrane reactor, Int.J.Hydrogen Energy xxx 1 -1 5

2011

DECAB: Process development of a phase change absorption process

E. Sanchez Fernandez, E. Kimball, E.L.V. Goetheer
TNO Separation Technology
Delft, the Netherlands

Introduction

This work describes a novel separation process to remove CO_2 from flue gas based on precipitating solvents (DECAB). The reaction of CO_2 with alkanol amines has been extensively described in the literature ([1], [2], [3], [4]). Depending on the type of amine, different species are predominantly formed. In the case of primary amines, the favored species is carbamate, as formed according to the equilibrium reaction (1).

$$2R-NH_2+CO_2 \leftrightarrow R-NHCOO^- + R-NH_3^+ \quad (1)$$

The process described here is an enhanced CO_2 absorption based on the Le Chatelier principle, which states that the equilibrium can be shifted by removing one of the constituents in the reaction. In reaction (1), the equilibrium can be shifted towards the formation of carbamate (R-$NHCOO^-$) by removing the protonated amine (R-NH_3^+) from the reaction medium, thus resulting in a higher absorption capacity. For this purpose an amino acid species has been selected. Amino acids react with CO_2 in principle in the same way as alkanol amines, as given by reaction (2).[5] However, amino acids are, in general, more stable than alkanol amines.

$$2\,^-OOC-C(R)-NH_2+CO_2 \leftrightarrow\, ^-OOC-C(R)-NHCOO^- +\, ^-OOC-C(R)-NH_3^+ \quad (2)$$

Unlike amines, amino acids form electrolytes in aqueous solutions that comprise a zwitterion. Due to its double charge, zwitterions have very limited solubility in water and can be easily separated from the medium by precipitation. In Figure 1, a hypothetical plot is given to depict the difference between a precipitation-based process and a conventional absorption process. This figure shows that for a given partial pressure of CO_2, a process with precipitation has higher loading than a process without precipitation. Therefore, the combined process (simultaneous absorption and precipitating process) leads to additional capacity of the solvent.

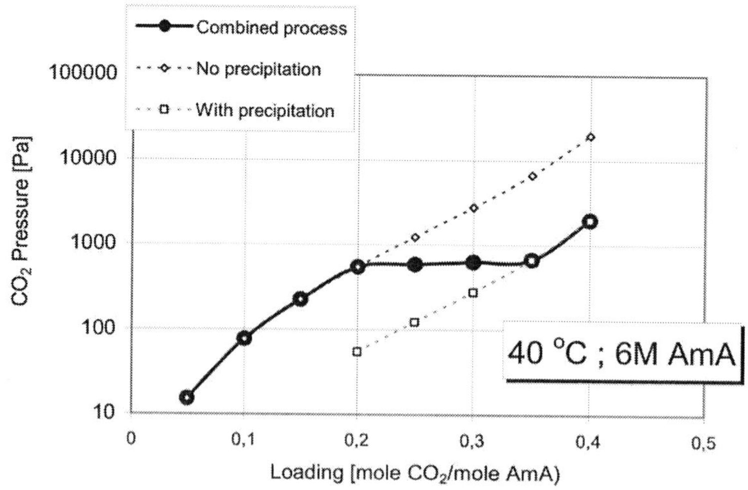

Figure 1 Schematic picture to depict the difference between a precipitating and a non-precipitating system in terms of CO_2 pressure as function of loading.

This works presents the conceptual design of a CO_2 capture process based on the above mentioned principle. The followed strategy combines the use of literature data, estimation of main properties, and a limited number of experiments. The first step in the design, described in Section 1, is the development of process concepts. This consists of selecting and ordering the necessary unit operations that will achieve CO_2 removal from flue gas and recycling of the solvent. Section 2 gives a description of the process and a compilation of technology options for the different unit operations. A structure is provided along with a ranking of options based on selected performance parameters. Section 3 gives a framework of key properties, thermodynamics, and kinetics that allow for the estimation of size and key performance parameters of the process. Based on shortcut methods, a conceptual design of the process is given in Section 4.

1 Process concepts development

The principle described in the previous section requires that a new capture process is designed. Figure 2 shows a block diagram that includes the main steps and functions of the process. The flue gas first needs to be contacted with the solvent (absorption block). The CO_2 then reacts with the solvent, as in reaction (2), resulting in a slurry that contains the carbamate and the precipitated amino acid. Crystals are dissolved (crystal dissolution block) and then the solvent is regenerated (Regeneration block). Finally, the resulting CO_2 product is compressed.

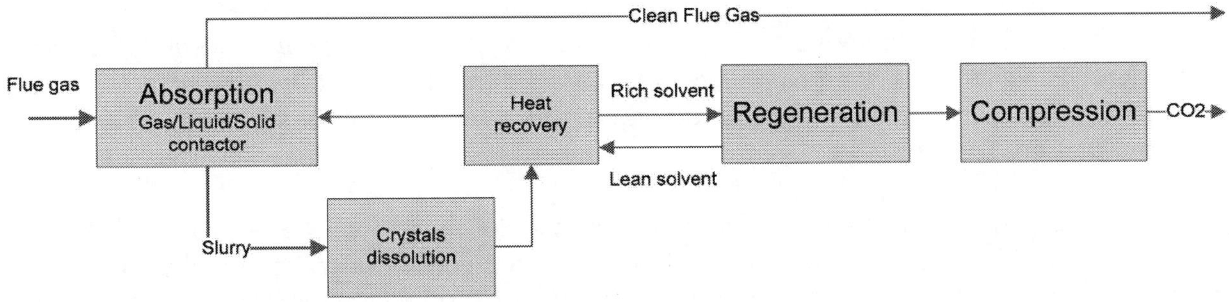

Figure 2 Block diagram showing the structure of the process

The process design allows for various degrees of freedom. The absorption process requires that the three phases (liquid/gas/solid) are contacted. Generally speaking, contactors can employ one of the following mechanisms: dispersion of a gas phase in a liquid or mixture of a liquid and a solid (bubbles), dispersion of liquid droplets in a gas phase (droplets), or contacting a gas phase and a liquid or slurry by means of a fixed contacting surface (surface). Based on these mechanisms, different contactors can be sought out and classified, as in Table 1. Moreover, the regeneration also offers several degrees of freedom (i.e. regeneration at low temperature and low pressure or regeneration at higher temperature and higher pressure).

For the selection of suitable contactors, the following criteria were considered:

1. Fouling behaviour: appropriate for slurry
2. Pressure drop: This parameter influences both process investment (CAPEX) and operational expenses (OPEX) but it has a more important effect in OPEX due to the power consumption to overcome the contactor's pressure drop.
3. Mass transfer characteristics: Mass transfer influences equipment size. This has a direct impact in process investment (CAPEX). In the present study, mass transfer of the different options is represented by the typical operational values of the overall mass transfer coefficient (K_{ol} [m/s])

4. Efficiency: Influences process investment and the purity of the clean flue gas leaving the contactor. This parameter is represented / related to the specific contact area (a $[m^2/m^3]$), Height of a theoretical plate (HETP) and number of transfer units (NTU).
5. Reliability: based on proven technology, applications in industry, size of operation.

Table 1 Generation of options and ranking for contacting systems. Rating key: **3- Best selection**, 2- Usually applicable, 1- Evaluate carefully, 0- Do not use

Contactor type	Contacting Mechanism[1]	a $[m^2/m^3]$	$K_{ol} \times 10^4$ [m/s]	Fouling	Pressure drop	Efficiency	Reliability	References
Random Packed column	Surface	100 - 200	0.5 - 2	1	2	2	2	[2],[6], [7], [8]
Structured Packed	Surface	250 - 500	0.5 - 2	1	2	**3**	2	[2], [6], [9]
Tray Column (dual flow)	Bubbles	150	1 - 4	2	1	2	2	[2],[6]
Spray column	Droplets	60 - 100	1 -3	**3**	**3**	0	2	[11],[12],[13],[14]
Multi Stage Crystallizer	Bubbles	20	1 - 4	2	1	2	2	[10]
Bubble column	Bubbles	20 - 400	1 - 4	**3**	1	2	2	[15],[16]
HiGee	Droplets	NA	NA	1	2	2	0	[19], [20]
Hollow fibre module	Surface	3000	0.1 – 0.5	0	1	2	1	[17], [18]

Table 1 shows quantitative estimates for the specific contact area and overall mass transfer coefficient. For the other criteria, the related parameters cannot be quantified at this stage. Instead, a qualitative grading is given based on literature review and the grading system given by Kohl and Nielsen, 1997.[6]

In the absorption case, fouling behaviour, efficiency, and pressure drop are the most important criteria. For the regeneration, efficiency is considered the most important criterion. It follows from this table that in the case of absorption there is no contactor that complies with both fouling and efficiency criteria. For this reason the absorption was divided into two steps and makes use of a spray tower (that handles the slurry when the solvent precipitates) and a packed column (that adds the necessary extra efficiency for 90% CO_2 removal).

In the case of regeneration the packed column is selected due to having a high efficiency, high volumetric ratio, and low pressure drop. In this case, fouling behaviour is relegated to lower importance.

2 Design framework

The recovery of CO_2 from flue gas based on solvent absorption is governed by the phase behaviour of CO_2 and the solvent at the conditions of the absorption unit. For most cases, temperature and pressure are used as operational variables in the regeneration unit in order to release the CO_2 and re-use the solvent. In the present case, the solvent also undergoes a phase change. In order to design the process, the phase behaviour of CO_2 needs to be known at the conditions of the absorption and regeneration units. Moreover, the key parameters influencing the phase change of the solvent (solubility, heat of crystallization, and slurry density) also need to be known. This information has been taken from literature, estimated, or measured.

[1] Surface stands for fixed contact area systems, Bubbles stands for dispersion of gas into a liquid or slurry and Droplets stands for dispersion of liquid into a gas

For the present case, the potassium salt of taurine was selected as the model solvent. The key information necessary for completing the design is listed below:

1. Pure components properties and mixture properties
2. Vapour-Liquid equilibrium of CO_2 with the solvent at absorption and regeneration temperatures
3. Vapour pressure of the solvent
4. Slurry density
5. Mass transfer coefficients realized for the selected contactor type

Table 2 Important pure component properties for Conceptual Process Design

Property	MW [g/mol]	T_m [°C]	T_b [°C]	ΔH_{vap} [kJ/mol]	ΔH_{dis} [kJ/mol]
Taurine	125	305.11	na	na	24.2
KOH	56	420	1327	na	na
Water	18	0	100	40.65	na

Table 2 shows the important pure components information. The Vapour-Liquid-Solid Equilibrium (VLSE) was measured for a 6M solution of potassium taurate. Figure 3 shows a plot of the VLSE of CO_2 at two different temperatures and a concentration of 6M. The same figure also shows the vapour pressure of the solvent mixture at different concentrations. Based on these experiments the solubility of CO_2 can be estimated. Also in these experiments, the precipitation behaviour can be observed and an estimate of the heat of absorption can be given (67kJ/mol CO_2). Sizing the contactors also requires information of equipment mass transfer and reaction kinetics. For shortcut calculations, the overall mass transfer coefficient was determined experimentally for a column packed with Mellapak 350Y and for a spray column under different gas to liquid ratios. Figure 4 shows the overall mass transfer coefficients for these contactor types.

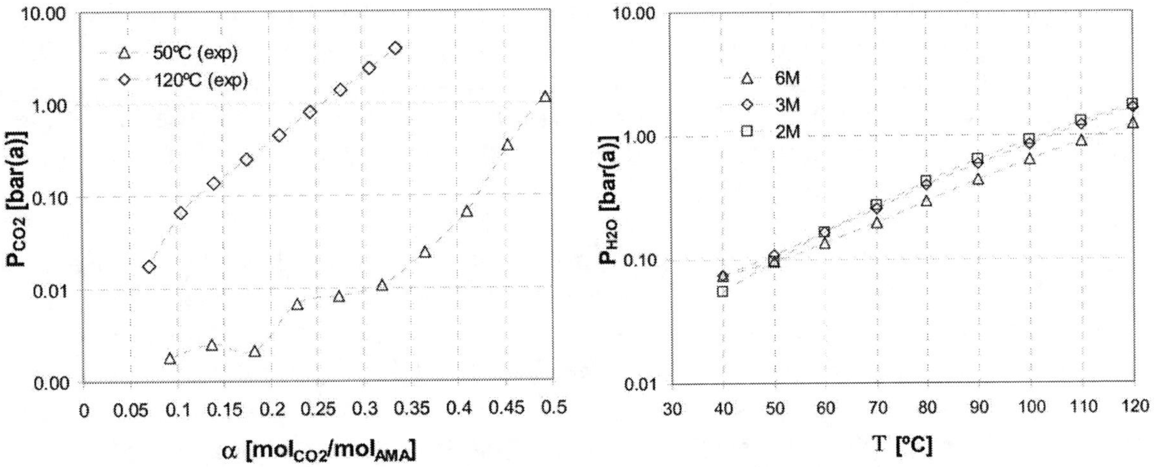

Figure 3 Left – Equilibrium behaviour of CO_2 in a 6M solution at 50ºC and 120ºC
Right – Water vapour pressure of solutions of potassium taurate of 2, 3, and 6M

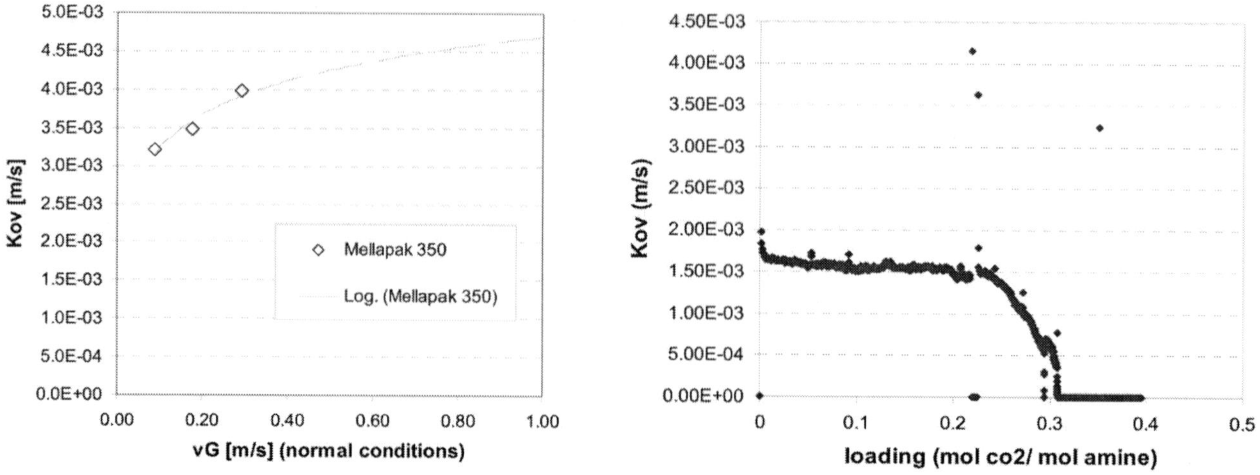

Figure 4 Left - Overall mass transfer coefficient for a column packed with Mellapak 350Y
Right - Overall mass transfer coefficient for a typical spray column experiment

3 Conceptual Process Design

This section presents a conceptual design of the process based on the information given in section 3. Figure 5 shows the process flow diagram of a process that treats refinery flue gas containing 8%vol CO_2 (170 °C, 1.06 bara) and achieve 90% capture. Initially, the flue gas is cooled in a quencher to the absorption temperature (50 °C). After cooling, the flue gas is contacted with pre-loaded solvent in a spray-tower, which consists of a column with no internals. The solvent is sprayed as fine droplets with a high surface area for contacting the gas with the solvent. As a result of this contact, the CO_2 undergoes a chemical reaction with the solvent that leads to the formation of carbamate and carbonate ions. As a consequence of this reaction, the pH of the solvent solution decreases and, with that, so does the solubility of the amino acid. When the solubility limit is reached, the amino-acid precipitates as an amino-acid zwitterion. The resulting slurry is collected at the bottom of the tower with a loading of $0.41mol_{CO2}/mol_{Taurate}$. The partial pressure of CO_2 in the flue gas is decreased until the point at which massive precipitation of the amino acid occurs. This point has been measured experimentally at different temperatures. Results show that at 50 °C and 6 M potassium taurate, this pressure is 4 kPa CO_2 (0.04 bar CO_2). The remaining CO_2 in the gas is captured in the absorption column, where the depleted flue gas is contacted with lean solvent. The absorption column is a conventional packed column filled in with Mellapak 350Y. There the CO_2 partial pressure is reduced to 0.8 kPa.

The assessment of the technical performance has been carried out by a process model implemented in a spreadsheet application. The aim was to give a rough estimation of the size of the equipment along with mass and energy balances. The most relevant design variables in the process are the temperature and pressure in the stripper (120°C and 1.4 bara) and the gas to liquid ratio in the absorber and spray-tower (542 STPvol/vol). The temperature and pressure in the stripper determine the degree of solvent regeneration and, therefore, the lean loading in the feed solution to the absorber (0.13 molCO_2/molTaurate).

These variables have a direct effect on the percentage of CO_2 removed and the specific energy consumption of the process. Several process calculations were performed until the optimal specific energy consumption was found. The relevant performance parameters corresponding to this point are listed in Table 4.

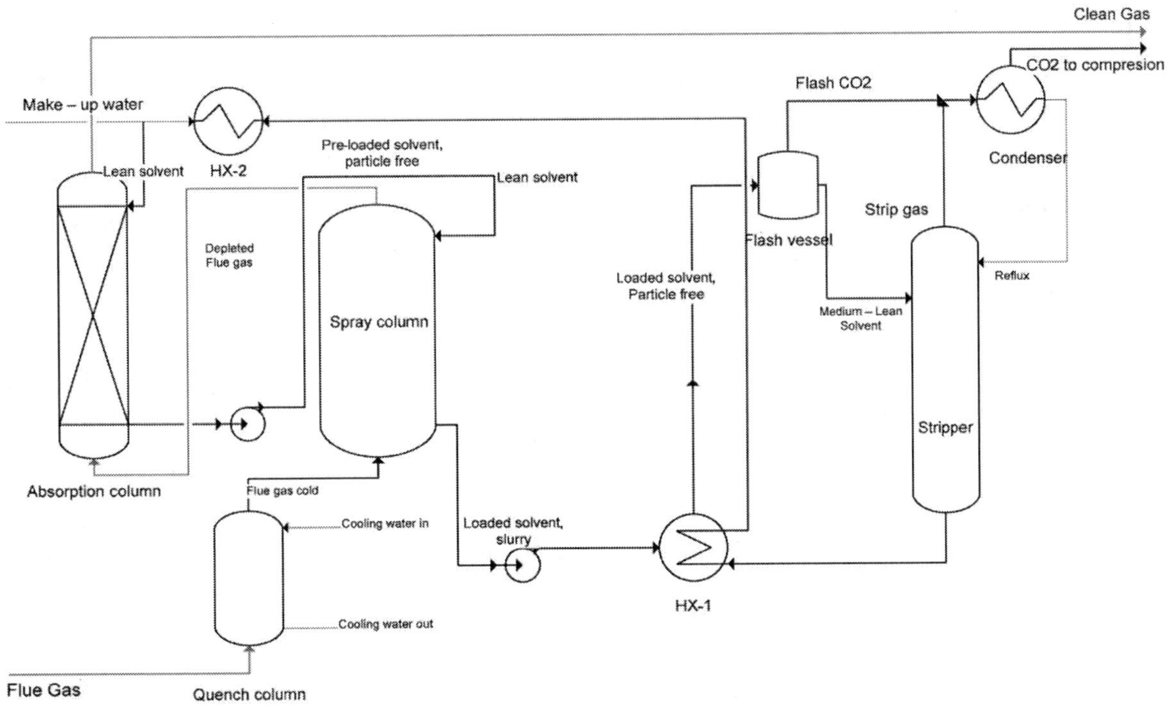

Figure 5 Process flow diagram for capturing 90% of the CO_2 from a refinery flue gas stream containing 8% CO_2

Table 4 Main performance parameters for the process

CO_2 Captured	98295	kg/h
Specific energy consumption	3.2	GJ/ton$_{CO2}$
Net loading	75	g/L
Contributions to the energy requirement		
Steam production	76135	kW
Crystals dissolution	9955	kW
Total energy	86090	kW
Cooling water requirements	133.11	m^3/ton$_{CO2}$
Make - up water	0.15	m^3/ton$_{CO2}$
Electricity	140.81	kWh/ton$_{CO2}$

The process described has a specific heat consumption of 3.2 GJ/ton$_{CO2}$. Part of this energy is necessary to dissolve the amino acid crystals and the other part is necessary to regenerate the solvent. However, the initial flue gas could be used to supply the heat duty for dissolving the crystals (10MW). This will result in a reduction of the specific duty to 2.8 GJ/ton$_{CO2}$. Considering the case of refinery flue gas, which contains less CO_2 than the flue gas from pulverized coal power plants, the given energy consumption is already lower than existing conventional processes.

It is evident that based on the above mentioned results, the DECAB principle can lower the energy consumption for CO_2 removal. This study has shown an example based on potassium taurate. However, by careful selection of appropriate amino acid solvent systems, significant energy reductions can be further achieved.

4 Conclusions

This work shows a process that uses precipitating solvents for CO_2 capture. A new process concept is given that is capable of handling slurries. The key advantages of this process are:

- It is environmentally friendly. Since the process is based on the salt of an amino acid, there are fewer emissions to the air and less degradation products.
- It has low energy consumption due to a lower vapor pressure of the solvent and higher net loading than conventional processes.
- It has the potential to significantly reduce energy consumption by exploring different process options for regeneration and selection of an appropriate amino acid solvent (i.e. with less crystallization energy).

The conceptual process developed allows for an economic evaluation and assessment of process options that will further lead to benefits over conventional processes.

5 References

[1] Danckwerts, P.V., The reaction of CO2 with ethanolamines, Chem. Eng. Sci. (34) 1979. 443-446

[2] Blauwhoff, P., Versteeg, G., Van Swaaij, W., A study on the reaction between CO2 and Alkanolamines in aqueous solutions, Chem. Eng. Sci., 39, 1984. 207-225.

[3] Versteeg, G.F., Duijk, van L.A.J. and Swaaij, van W.P.M., On the kinetics between CO2 and Alkanolamines both in aqueous and non-aqueous solutions. An overview, Chem. Eng. Comm. (144) 1996. 113-158

[4] Barth, D.,Tondre, C., Delppuech, J.-J., Kinetics and mechanisms of the reactions of carbon dioxide with alkanolamines: a discussion concerning the cases of MDEA and DEA, Chem. Eng. Sci. (39) 1984. 1753-1757

[5] Kumar, P.S., Hogendoorn, J.A., Versteeg, G.F. and Feron, P.H.M.., Kinetics of the reaction of CO2 with aqueous potassium salt of taurine and glycine, AIChE Journal. (49) 2003. 203-213.

[6] Kohl A., Nielsen R., Gas purification, Gulf chemical Engineering. 1997 Fifth edition

[7] Metal Random Packing. Sulzer Chemtech specifications

[8] Structured Packings. Sulzer Chemtech specifications

[9] Groserichter, D. and Stichlmair, J., Crystallization fouling in packed Columns, Trans IChemE (81) Part A 2003. 68 - 73

[10] Thieme, C., Sodium carbonates, Ullmann's Encyclopedia 2005

[11] Kuntz, J. and Aroonwilas A., Performance of spray column for CO2 capture application, Ind. Eng. Chem. Res (47) 2008. 145-153.

[12] Warych J. and Szymanowski M., Model of the wet limestone flue gas desulfurization process for cost optimization, Ind. Eng. Chem. Res (40) 12 2001. 2597-2605.

[13] Wijlhuizen A.E., Kerkhof P.J.A.M. and Bruin S., Theoretical study of the inactivation of phosphatase during spray drying of skim-milk, Chem. Eng. Sci (34) 1979 651-660

[14] Wirsching F., Calcium Sulfate, Ullmann's Encyclopedia 2005

[15] Thorat B., Kataria, K., Kulkarni, A.V. and Joshi J.B., Pressure Drop Studies in Bubble columns, Ind. Eng. Chem. Res (40) 2001. 3675-3688.

[16] van't Riet K., Basic Bioreactor Design, Marcel Dekker New York 1991

[17] Zhang, X., Zhang, P., Wei, K., Wang, Y. and Ma, R., The study of continuous membrane crystallization on lysozyme, Desal. (219)2008 101-117

[18] Drioli E., Membrane contactors: fundamentals, Applications and potentialities, Membrane science and technology series. Elsevier (2006)

[19] Lin, C.-C., Liu, W.-T. and Tan, C.-S., Removal of Carbon dioxide by Absorption in a rotating packed bed, Ind. Eng. Chem. Res (42) 11 2003. 2381-2386

[20] Tan C.S. and Chen J.E., Absorption of carbon dioxide with piperazine and its mixtures in a rotating packed bed, Sep. Pur. Technol. (49) 2 2006. 174-1

(92c) A Graphical Method of Improving Batch Reactors

David Ming, Centre of Material and Process Synthesis (COMPS), University of the Witwatersrand, Johannesburg, South Africa, **Diane Hildebrandt**, Center of Material and Process Synthesis, University of the Witwatersrand, Johannesburg, South Africa, Johannesburg, South Africa and David Glasser, Centre of Material and Process Synthesis, University of the Witwatersrand, Johannesburg, South Africa

A graphical technique has been used to determine the production efficiency of batch reactors. We find that by constructing line segments in concentration—time space, between the initial and final states in a standard batch reactor we may determine the production rate directly from the gradient of the line. The gradient of this line is found to be inversely proportional to the production rate corresponding to the batch and hence the batch performance is maximised when the abovementioned gradient approaches its smallest value possible.

We find that interpreting production rate in this manner has identified a new opportunity for improving performance. The opportunity results from the discovery that mixing allows one to overcome areas of low production. As a consequence, the modification is a method of potentially improving production rate in batch reactors by mixing final (desired) product with feed material. The resulting modification shows an improvement over the regular method of operation when no mixing is used when the kinetics is such that the shape of the associated batch profile is concave in concentration—time space. The method is illustrated with industrially significant examples.

(92e) Modeling of An Autothermal Catalytic Monolith Reformer to Obtain Hydrogen for Fuel Cells

Safdarnejad Seyed Mostafa and Mohammad Hassan Eikani, Green Research Center and Chemical Engineering Department, Iran University of Science and Technology (IUST), Tehran, Iran

Autothermal reforming (ATR), that is the combination of partial oxidation and steam reforming, is the preferred method for transportation applications because it is a self-sustained process and as a result, increases the energy conversion efficiency and correspondingly reduces environmental impact. The main scope of this work is to simulate the performance of an autothermal catalytic monolith reformer for the production of hydrogen, which is intended to be used in a fuel cell. A three dimensional numerical model was developed based on the computational fluid dynamics (CFD). In the model of monolithic reformer, a single channel was used assuming that each channel behaved the same. The catalysts were deposited on channel walls of the monolith and assumed that the chemical reactions only happened on the catalytic wall surface. The mass transport, chemical reactions and heat transfer can be analyzed simultaneously in the single channel with the aid of CFD simulation. The correctness of the model was verified by comparing the simulation results with the experimental data available in literatures.

(92f) Investigation of Traditional and Novel Techniques for Cooling In High Heat Flux Microelectronics Applications

Julaunica Tigner, Mechanical Engineering, Tuskegee University, Tuskegee, AL and Tamara Floyd-Smith, Chemical Engineering Department, Tuskegee University, Tuskegee, AL

The growing demand for electronic devices to be smaller and faster has increased the energy released by these devices in the form of heat. Devices such as laptop computers are not exempted from these demands. The primary traditional technologies currently used to remove heat generated in these devices are fins and fans. Fins dissipate heat through contact with the surface area of a thermally conductive material like aluminum or copper. Fans dissipate heat through air convection. The objective of this study is to perform benchmarking experiments of traditional technologies and to compare the results with more novel technologies like the use of micro channels with water as the cooling fluid. Results are forthcoming.

Catalytic membrane reactors for environmental applications

Anne Giroir-Fendler

Université de Lyon, Lyon, F-69003, France, Université Lyon 1, IRCELYON UMR 5256 CNRS- UCBL1, 2 avenue Albert Einstein, Villeurbanne, F-69622, France.

Application of membrane reactor can improve the conversion of the catalytic processes considerably. In a wide range of applications and more particularly for environmental application, significant progresses have being made on membrane development. The feasibility on laboratory scale has been demonstrated for membrane reactor technology since 20 years. In this paper, we will present some catalytic membrane reactors development devoted to environmental application like VOCs abatement, waste water treatment, gas separation and xylene isomerisation.

VOC abatement - The destruction of VOCs is a great challenge and also of great industrial importance. Our aim is to investigate the catalytic oxidation of propene and toluene in a membrane reactor used in a flow-through contactor configuration. In this case, the gas mixture of reactants is forced to go through the membrane, i.e. through the catalytic pores. This study on propene and toluene combustion showed that the catalytic membrane reactor performed better than the conventional monolithic reactor in term of efficiency [1]. This confirms that membrane reactor is a promising alternative for the combustion of VOCs and that the flow-through membrane reactor may lead to decreased light-off and total VOC combustion temperature with a low overall Pt loading. In order to increase the efficiency of the catalytic phase, new routes of platinum deposition have been investigated and showed good results in terms of stability and catalytic performances [2]

Waste water treatment - Wet air oxidation (WAO), earlier developed by Zimmerman during the 1950s, constitutes an attractive technology for the treatment of industrial effluents containing low to intermediate concentrations of refractory and toxic compounds for which incineration or biological remediation are inefficient. The efficiency of WAO can be improved by the use of heterogeneous catalysts. In order to improve the gas/liquid/solid contact in CWAO, the development of innovative catalytic reactors is desired. The use of catalytic membrane reactors (contactor-type), coupling a catalyst and a membrane in the same unit, could be an option. Also in this configuration, a separation of the catalyst from the product is not required as the catalyst is loaded as nanoparticles in the porous media of the membrane structure. In this study, we have prepared catalytic membrane using several techniques of deposition in order to well control the position of the active phase in the porous structure. After optimization of the experimental parameters, the study of pollutants degradation has showed that catalytic membrane reactor, in contactor configuration present highest efficiency than conventional reactor due to optimized contacts between reactants and active sites [3].

Gas separation - Zeolite membranes are the good candidate to separate chemical products and in this study we focused on the preparation of highly reproducible nanocomposite MFI hollow fibers membranes via pore-plugging, and on the applications of CO_2/N_2 and p-xylene/m-xylene separations [4]. In order to increase the separation factors of these materials, nanocomposite B-MFI hollow fibers were also successfully synthesized in one pot by substitution of Si with B during synthesis. B-MFI hollow fibers improved the separation performance compared to MFI membrane on the vapor separation of xylene and n-hexane isomers [5]. A modeling study on n-hexane/2,2-dimethylbutane separation reflects that hexane separation within both MFI and B-MFI hollow fiber membranes is mainly driven by diffusion differences and configurationally entropy effects [6].

Xylene isomerisation - This part presents isomerisation of meta-xylene over Pt-HZSM-5 in an extractor-type catalytic membrane reactor (e-CMR) using tubular nanocomposite-alumina MFI membrane. The membrane used for the study was prepared via a pore-plugging synthesis technique and it showed p-xylene permeance of 11.4nmol/(m^2.s.Pa) [7]. Meta-xylene saturated in N_2 was fed into the membrane and N_2 gas swept over the outer surface of the membrane (for e-CMR). The performance of the e-CMR was compared with that of a fixed bed reactor (FBR) with similar reactor dimensions and operation conditions. Thus e-CMR showed 34.5% increase in para-xylene selectivity when compared to FBR at 400°C and these results are in good agreement with the literature [8].

[1] S. Bénard, A. Giroir-Fendler, P. Vernoux, N. Guilhaume and K. Fiaty, Catalysis Today, Volume: 156 Issues 3-4, Pages: 301-305, 2010

[2] Yacou, A. Ayral, A. Giroir-Fendler, A. Baylet and A. Julbe, Catalysis Today, Volume: 156 Issues 3-4, Pages: 216-222, 2010

[3] M. Alame, A. Abusaloua, M. Pera-Titus, N. Guilhaume, K. Fiaty, A. Giroir-Fendler, Catalysis Today, Volume: 157 Issues 1-4, Pages: 327-333, 2010

[4] Z. Deng, C-H. Nicolas, M.O. Daramola, J. Sublet, Th. chiestel, A.J.Burger, Y. Guo, A. Giroir-Fendler and M. Pera-Titus, Journal of Membrane Science, Volume: 364 Pages: 18-26, 2010

[5] Z. Deng, C-H. Nicolas, Y. Guo, A. Giroir-Fendler and M. Pera-Titus, Microporous and Mesoporous Materials, Volume: 133 Issue 1-3 pages 18-26, 2010

[6] Z. Deng, M. Pera-Titus, Y. Guo and A. Giroir-Fendler, Industrial and Engineering Chemistry Research, Volume 49, pages 11697-11707, 2010

[7] M.O. Daramola, Z. Deng, M. Pera-Titus, A. Giroir-Fendler, S. Miachon, A.J. Burger, L. Lorenzen & Y. Guo, Catalysis Today, Volume: 156 Pages: 261-267, 2010

[8] M.O. Daramola, A.J. Burger, A. Giroir-Fendler, S. Miachon, L. Lorenzen, Applied Catalysis A: General, Volume: 386, Issue: 1-2, Pages: 109-115, 2010

(95b) Bacterial Community Structure In a Hydrogen-Based Membrane Biofilm Reactor

Liang Duan[1], Yonghui Song[1], Siqing Xia[2] and Slawomir W. Hermanowicz[3], (1)Section of Urban Water Environmental Research, Chinese Research Academy of Environmental Sciences, Beijing, China, (2)Department of Environmental Science and Engineering, Tongji University, Shanghai, China, (3) Department of Civil and Environmental Engineering, University of California, Berkeley, CA

Abstract. The whole bacterial community structure in a novel hydrogen-based membrane biofilm reactor (MBfR) was clarified by PCR-DGGE and high-density universal 16S rRNA Microarray, which includes 506,944 probes targeted to 8935 clusters in 16S rRNA gene sequences. The PCR-DGGE results shown bacterial communities evolved with operation time in MBfR. It was obviously that the bacterial number increased as time increased. The Microarray results shown 360 and 997 bacterial subfamilies were detected in MBfR operated at 10 days and 40days. The whole bacteria can separate as 19 and 24 phyla in two samples, respectively. Proteobacteria was the predominant phylum, 270 and 505 subfamilies were found in two samples. Alphaproteobacteria was the dominant classes in Proteobacteria and occupied 39.3% and 45.5% in two samples, respectively. Acidobacteria, Firmicutes and Bacteroidetes were the subdominant groups, occupying around 14.2%, 11.2% and 11.2% in MBfR at stable stage. Some bacteria groups included Cyanobacteria, Verrucomicrobia, Planctomycetes, Chloroflexi and Spirochaetes phylum also found more than 15 subfamilies. All the results shown the MBfR system had more bacterial community diversity in stable stage. Moreover, the phylogenetic trees of top 50 bacteria were constructed. It were selected by compared the relative average fluorescence intensity of the whole bacteria in MBfR.

(95c) Catalytic Membrane Reactor In the Treatment of Waste Water

Brenda Martínez[1], **Rodriguez Miguel Torres**[1], Mirella Gutiérrez Arzaluz[1], Montoya De la fuente Ascencion[2] and Julia Aguilar Pliego[1], (1)Ciencias Básicas, Universidad Autonoma Metropolitana, Mexico, Mexico, (2)Ing Molecular, Universidad Autonoma Metropolitana, Mexico, Mexico

Catalytic membrane reactor in the treatment of waste water

Martínez. B[1], Torres. M[1]*, Montoya. A[2], Gutiérrez. M[1], Aguilar. J[1]

[1]*Departamento de Ciencias Básicas, Universidad Autónoma Metropolitana-Azcapotzalco, Av. San Pablo Nº 180, Col. Reynosa Tamaulipas, Azcapotzalco, 02200, México, D.F*
[2]*Direccion de Investigación y Posgrado, Instituto Mexicano del Petróleo, Eje Central Lázaro Cárdenas Norte Nº 152, Col. San Bartolo Atepehuacán, 07730, México, D. F.*
**e-mail trm@correo.azc.uam.mx*

Keyword, Membrane reactor, Catalytic wet oxidation, Nanoparticles, Contactor

A membrane reactor with a liquid and gas flowing on different sides of the catalytic membrane is called catalytic diffuser. The gas diffuses through the macropores membrane layer and dissolves in the waste water at the interface.

In this work we use a diffuser membrane reactor configuration for the elimination of formaldehyde in water. The active phase of catalytic membrane is gold nanoparticles that are dispersed onto the macropores of the membrane. The ceramic support is a tubular membrane of the company Pall-Exekia (α-alumina, T1-70), with dimensions: 15 cm length, inner diameter of 0.7 cm and outer diameter of 1.0 cm, with an average pore size of 0.2 microns. The catalytic test consists of a feed zone, a reaction zone and analysis zone. The catalytic membrane is mounted on a stainless steel cylinder jacket containing the ceramic membrane, this configuration was divided the reactor in two chambers, the liquid is fed to the inner membrane with a pump and the outer one was used for feeding the pure oxygen gas-phase, it imposed a differential pressure between the two phases of 0.5 bar, and the oxygen pressure was 5 bar. The oxidation of formaldehyde was studied at a temperature of 60 ° C with three different concentrations of formaldehyde, 200, 300, 500 ppm. In a time of 60 min was reached a conversion of 95% for concentrations of 100 and 300 ppm and 38% for 500 ppm. The reaction products were determined by GC. The catalytic membrane reactor shows good performance for this reaction.

Escucha
Leer fonéticamente

Diccionario - Ver diccionario detallado

Membrane reactors for PEM based micro-cogeneration systems: System assessment

Fausto Gallucci[1,*], Martin van Sint Annaland[1], Leonardo Roses[2], Giampaolo Manzolini[2], Stefano Campanari[2]

[1] Chemical Process Intensification, Chemical Engineering and Chemistry, Eindhoven University of Technology, Eindhoven, The Netherlands
[2]Politecnico di Milano, Dipartimento di Energia, Via Lambruschini 4, 20156 Milano, Italy

(*) corresponding author. Tel: +31 40 247 3675, e-mail: f.gallucci@tue.nl

Abstract

Nowadays a great part of the scientific community agrees that anthropogenic CO_2 emissions need to be reduced significantly in the near future to prevent excessive global warming. A possible solution to this problem is the use of advanced devices able to increase the energy efficiency. In this respect PEM fuel cell based cogeneration systems are potential solutions for further improvement of the energy efficiency. These devices will most likely be fed with hydrogen produced by reforming of light hydrocarbons such as natural gas. Previous studies have shown the advantages of palladium-based membrane reactors (MR) for ultra-pure hydrogen production over conventional natural gas processors for feeding the fuel cell.

In this work, both experimental and simulation studies were carried out to quantify the performance of different technological MR approaches such as fluidized bed membrane reactors and packed bed membrane reactors. In contrast with the packed bed membrane reactors, membrane-assisted fluidized bed reactors offer the possibility to operate under virtually uniform temperature conditions, which is beneficial for the stability of membranes, but also require lower membrane areas for the same performance because concentration polarization is avoided. The experiments have shown that the hydrogen flux through the membranes and the perm-selectivity of the membranes are stable during two weeks continuous operation under bubbling fluidization conditions.

Simulations have shown pronounced axial temperature profiles in packed bed membrane reactors, which are responsible for an increase of up to 25% in the membrane area required in comparison to fluidized bed membrane reactors. In all the cases, a similar total efficiency can be achieved that is 9% higher than the total efficiency of a conventional reformer/fuel cell system (however with different required membrane areas for the packed bed and fluidized bed membrane reactors). A parametric study has been carried out and the possibilities to integrate CO_2 capture inside the PEM micro-cogeneration system have been evaluated.

Keywords: Membrane reactors, Membrane reformer, fluidized bed, micro-cogeneration, hydrogen production, Palladium membrane

Octane Boosting in a Membrane Assisted Radial Flow Naphtha Reactor, Using DE Optimization Technique

M.R. Rahimpour[1], D. Iranshahi, E. Pourazadi

School of Chemical and Petroleum Engineering, Department of Chemical Engineering,

Shiraz University, P.O. Box 71345, Shiraz, Iran

Abstract:

The significant impact of catalytic naphtha reactors on the total benefits from the refinery complexes has been road toward a continuous evolution and significant advances in naphtha reactors and its operational conditions. Regarding this, the present study proposes a tubular membrane reactor with the radial-flow patterns of the sweeping gas and the naphtha feed named RF-TMR as a novel configuration instead of a conventional axial-flow tubular packed bed naphtha reforming process. The cross section area of the reactor is divided into some subsections. The walls of the gaps are coated by a Pd-Ag membrane layer to increase the hydrogen permeation rate. The membrane layer enhances pure hydrogen while boosts the octane number of reformates by increasing the aromatic production. The DE optimization technique is utilized here to find proper operational and dimensional parameters. Operating pressures (reaction and permeation sides), catalyst mass distribution in each reactor, total number of subsections, the ratio of length to diameter (LOD) for each reactor and the proper ratio of sweep to catalytic side angles are considered as the most important decision variables. The results show successful aromatic production in addition an increase about 27 kmol/hr in the hydrogen production rate compared with the conventional naphtha reactor. The idea is capable to apply

[1] Corresponding author. Tel.: +98 711 2303071; fax: +98 711 6287294.

E-mail address: rahimpor@shirazu.ac.ir (M.R. Rahimpour).

and revamp the fixed and moving bed radial-flow naphtha reactors. Mathematical modeling and also optimization technique should be apply for such a new ideas before commercialization and making decision for set up of a new pilot plants.

Keywords: Chemical Reactions; Membrane concept; Radial flow reactor; catalytic naphtha reforming; Octane boosting; Hydrogen production.

Conventional Axial-flow tubular packed bed reactor (CTR)

A simplified schematic of CTR is explained in Fig.1.

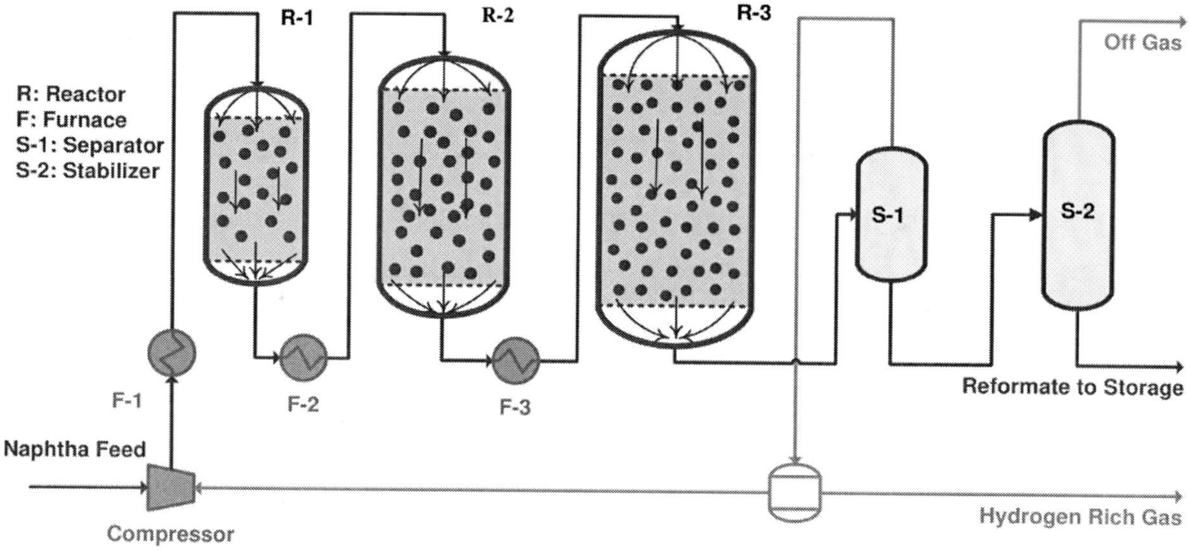

Figure.1

The naphtha feed is mixed with the recycled gas containing 60-90% (by mole) hydrogen and preheated before entering the 1st reactor. Reactors are packed with catalysts and the chemical reactions take place on catalysts' surfaces. Since naphtha reforming is an endothermic process, the outlet stream must be preheated before entering the following reactor by inter-stage furnaces. In order to stabilize the liquid and separate the gaseous product, the effluent from the 3rd reactor is cooled and directed into the separators. The liquid product is called reformate which mainly consists of aromatics (60–70 mass% of naphtha feed) and saturates in the C_5–C_9 carbon range. The main reactions in the first reactor are dehydrogenation and isomerization, in the second reactor are dehydrogenation, isomerization, cracking and dehydrocyclization and in the third one are cracking and dehydrocyclization [1].

Conceptual for proposed configuration (RE-TMR)

A schematic of proposed configuration is illustrated in Fig.2.

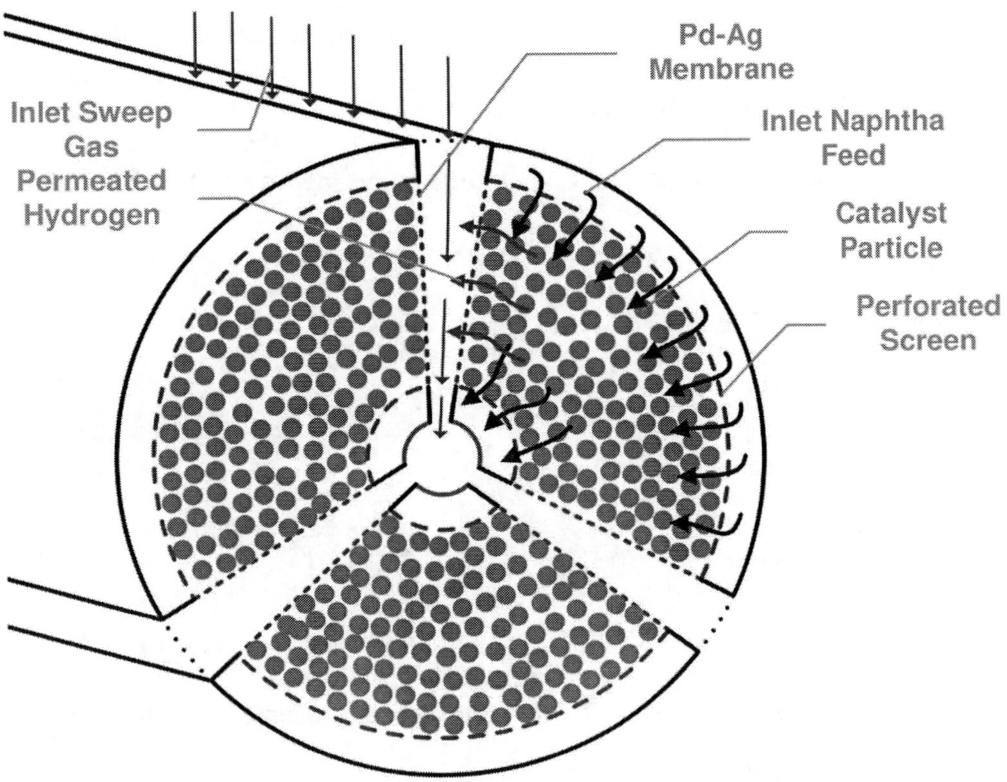

Figure.2

The internal design is such that the flow enters the vessel across the entire cross section of the vessel, first proceeding downwards near the wall of the vessel and then radially inwards, through the bed of catalytic particles and finally in a downwards direction via an axial collector. The reformed naphtha is collected by the axial collector pipe. The walls of the gaps between these subsections are coated by the Pd-Ag membrane layer, thus hydrogen permeates through the membrane layer and enters the shell side. Moreover, the sweeping gas flows radially in the shell side and carries the permeated hydrogen. Due to negligible pressure drop, smaller catalysts' particles with minor internal mass transfer resistance can be used.

In order to investigate the reactor behavior, four dominant reactions are considered based on the Smith's model to predict the outlet compositions of paraffin, naphthene and aromatics which are known as PNA. It is valuable to mention that a dynamic homogeneous one-dimensional model is used. All the physical properties are considered to be variable along the reactor. Orthogonal collocation method is used, as a powerful method in solving the PDEs, to achieve dynamic results of modeling. The validity of model is examined by comparison of daily report of conventional process during 800 days of operation.

Optimization results

Differential evolution (DE) is a heuristic approach based on optimization evolutionary algorithms and random search methods. This technique was presented by Storn and Price (1996) [2]. A Complete list of optimized parameters with their values is presented in Table 1.

Table 5 – Values of optimized parameters.

No.	Decision variable	Min.	Max.	Optimized value
1	Sweep gas pressure for the 1st reactor (kPa)	102	1000	592
2	Sweep gas pressure for the 2nd reactor (kPa)	102	1000	781
3	Sweep gas pressure for the 3rd reactor (kPa)	102	1000	177
4	Membrane thickness of the 1st reactor (μm)	0.000001	0.000030	0.000028
5	Membrane thickness of the 2nd reactor (μm)	0.000001	0.000030	0.000030
6	Membrane thickness of the 3rd reactor (μm)	0.000001	0.000030	0.000025
7	Total molar flow rate of sweep gas (kmol/hr)	100	1000	997.2
8	Inner radius of collector for the 1st reactor (m)	0.10	0.50	0.05
9	Inner radius of collector for the 2nd reactor (m)	0.10	0.50	0.05
10	Inner radius of collector for the 3rd reactor (m)	0.10	0.50	0.05
11	Compressor discharge pressure to the 1st reactor (kPa)	2000	3703	3697
12	Catalyst mass distribution for the 1st reactor (weight fraction)	0.01	0.99	0.17
13	Catalyst mass distribution for the 2nd reactor (weight fraction)	0.01	0.99	0.23
14	Catalyst mass distribution for the 3rd reactor (weight fraction)	0.01	0.99	0.60
15	Hydrogen mole fraction in the sweep gas	0	0.99	0.89
16	Hydrogen mole fraction in the recycle gas	0.40	0.75	0.50
17	LOD of the 1st reactor	3	5	3.84
18	LOD of the 2nd reactor	3	5	3.50
19	LOD of the 3rd reactor	3	5	4.50
20	The fraction of the total sweep gas to the 1st reactor	0.01	0.99	0.20
21	The fraction of the total sweep gas to the 2nd reactor	0.01	0.99	0.40

22	The fraction of the total sweep gas to the 3rd reactor	0.01	0.99	0.40
23	Number of sweep sections (NOS) for the 1st reactor	2	10	3
24	Number of sweep sections (NOS) for the 2nd reactor	2	10	3
25	Number of sweep sections (NOS) for the 3rd reactor	2	10	5
26	The ratio of permeation angle to the reaction one's ($\frac{\theta_1}{\theta_2}$) for the 1st reactor	0.01	0.25	0.05
27	The ratio of permeation angle to the reaction one's ($\frac{\theta_1}{\theta_2}$) for the 2nd reactor	0.01	0.25	0.05
28	The ratio of permeation angle to the reaction one's ($\frac{\theta_1}{\theta_2}$) for the 3rd reactor	0.01	0.25	0.05
29	Total fresh naphtha freed to the 1st reactor	266.4	350	276.24

Lower operational pressure in the permeation side is better from the economical point of view. However, it is worth mentioning that reducing the sweep gas pressure provides a stronger driving force for hydrogen permeation (lower H_2/HC) which in turn causes rapid catalysts deactivation. Membranes thickness is a key parameter for hydrogen permeation. Thickness reduction means less resistance for hydrogen permeation. However, as previously discussed, only the excessive amount of hydrogen needed for catalyst deactivation prevention ($H_2/HC=4.73$) must be removed from the reaction side and therefore it is not possible to reduce the thickness beyond a certain value and an optimum value should be determined. Sweep gas molar flow rate is also investigated in this work. Since the naphtha reforming reaction is endothermic, the temperature decreases along the reactor bed which is undesirable and it should be avoided. Therefore, the catalysts are distributed in three consecutive reactors with intermediate furnaces. In this work, the optimum catalysts mass distribution for each reactor is obtained. The hydrogen mole fraction in the sweep gas stream needs to be optimized. Higher hydrogen content of sweep gas reduces the need for high operating pressure and related cost .In this work, the value of hydrogen mole fraction in the recycle stream is optmized. Increasing the number of subsections provide higher heat transfer area to the reaction side and increases the products' yields. On the other hand, more subsections cause more hydrogen extraction from the reaction side and reduces the H_2/HC. Consequently the proper number of gaps should be defined. $\frac{\theta_1}{\theta_2}$ ratio is also optimized for all three reactors. Finally the fraction of sweep gas allocated to each reactor is optimized.

According to the achieved results from the optimization step, hydrogen and aromatic production rates are increases by 10.0 and 5.4 % compared with the non-optimized RF-TMR. The optimized reactor is capable to increase the reformate production rate by 3.9 % more than the CTR and 1.62 % boosting in aromatic content.

References:

[1] Khosravanipour Mostafazadeh A, Rahimpour MR. A membrane catalytic bed concept for naphtha reforming in the presence of catalyst deactivation. Chem Eng Process 2009; 48: 683–94.

[2] Storn R, Price K, Differential Evolution – A Simple and Efficient Heuristic for Global Optimization over Continuous Spaces. J Global Optim 1997; 11: 341–359.

(95f) Membrane Reactor Methanol Reformer

Gökhan Alptekin, Matt Cates and Michael Bonnema, TDA Research Inc., Wheat Ridge, CO

PEM fuel cells are potentially highly efficient power sources for soldier-portable power systems and other military power production equipment, as well as for many consumer devices a major drawback that prevents the widespread use of PEM fuel cell technology is the lack of compact units to store the hydrogen gas essential for their operation. One way to avoid hydrogen storage is to produce hydrogen on demand using compact fuel processors. Sulfur-free fuels, such as methanol offer significant advantages, simplifying the fuel processing. Methanol can be reformed at low temperatures. However, the hydrogen produced by fuel reforming will always be contaminated with diluents (such as nitrogen or CO_2) and a major anode electro-catalyst poison (CO). It is necessary to remove CO from the hydrogen feed to ensure the long-term stable electrochemical operation of the fuel cell.

A membrane reactor-reformer to provide a high purity hydrogen fuel that could be readily used by fuel cells was developed. THis system integrates the membrane steam-reformer module with a highly effective methanol reforming catalyst to combine the hydrogen production and gas processing steps into one process. Both the new catalyst and the membrane reduce the amount of water to be stored by achieving sub-stoichiometric reforming of methanol. This demonstrates the methanol-membrane reformer in a breadboard system; this showed the performance of the catalyst and the membrane integrated in the same module to produce high purity hydrogen through several heat-up/ cool down cycles and 72 hour missions with no degradation in the performance of the membrane or the catalyst. In this paper, we will present the results of the demonstration tests.

(95g) Investigation of the Naphtha Reforming Process and Hydrodealkylation of Toluene In a Novel Optimized Membrane Coupled Reactor

Vida Meidanshahi, **Ali Mohammad Bahmanpour**, Mohammad Reza Rahimpour and Davood Iranshahi, School of Chemical & petroleum Engineering, Shiraz University, Shiraz, Iran

In the current research, the conventional naphtha reforming reactors have been replaced by the thermally coupled reactors. The tube wall and the shell wall of a thermally-coupled reactor containing hydrodealkylation (HDA) of toluene in the exothermic side and naphtha reforming process in the endothermic side have been substituted by hydrogen-permselective membrane layers. A part of the produced hydrogen in the shell side enters the exothermic side to provide the necessary hydrogen supply for the HDA process. Permeated hydrogen through the shell wall enters another section called permeation side to be collected by sweep gas. This three-sided reactor has been optimized using differential evolution (DE) method to predict the operating conditions at which the conversion of the toluene and aromatics and also the hydrogen recovery yield would be maximized. The initial molar flow rates of the exothermic and permeation sides, as well as the inlet pressure and the composition of the sweep gas, plus the inlet temperature of the exothermic side are among the main decision variables in this research. The conversion of toluene and aromatics in the mentioned membrane reactor has been boosted significantly in comparison with the conventional reactors.

(95h) Perovskite Microtubular Membranes for Hydrogen Production From Water Splitting

Rafael V. Franca, School of Chemical Engineering and Advanced Materials, Newcastle University, Newcastle-upon-Tyne, United Kingdom, Alan Thursfield, Chemical Engineering and Advanced Materials, University of Newcastle, Newcastle-Upon-Tyne, United Kingdom and Ian S. Metcalfe, Chemical Engineering and Advanced Materials, University of Newcastle, Newcastle-upon-Tyne, United Kingdom

Here we present promising results for pure hydrogen production by chemically-driven water splitting using perovskite membranes. Steam is fed to one side of the oxygen ion and electron conducting perovskite membrane at high temperature (900°C). The water molecule dissociates on the membrane surface producing oxygen and hydrogen. The oxygen is transported through the membrane, as oxygen ions via mobile oxygen vacancies, shifting equilibrium in the direction of dissociation. The oxygen ions are then removed on the other side of the membrane using a reducing sweep gas. The reaction of this reducing gas with water provides the overall driving force for reaction and permeation. This project aims to develop and characterise hollow fibre perovskite membranes ($La_{0.6}Sr_{0.4}Co_{0.2}Fe_{0.8}O_{3-\delta}$) and subsequently test them for hydrogen production by the water splitting reaction using methane as a reducing gas. The overall reaction is the steam reforming of methane.

$La_{0.6}Sr_{0.4}Co_{0.2}Fe_{0.8}O_{3-\delta}$ (LSCF6428) hollow fibre membranes of 1.5 mm external diameter and 250µm wall thickness were prepared by phase inversion followed by sintering at high temperature (1300°C). Initially, an oxygen permeation experiment was carried out in order to test the oxygen flux through the hollow-fibres; as expected the oxygen permeation rates had a dependence on temperature and sweep gas flow-rates. Membrane-based hydrogen production experiments were carried out after successfully demonstrating that the membranes could permeate oxygen. From the membrane-based hydrogen production experiment, it was found that hydrogen could be produced continuously by membrane-based steam reforming for more than 250 hours, demonstrating high stability of the membranes under reaction conditions. At the same time as hydrogen was produced by water splitting on one side of the membrane, hydrogen, carbon dioxide and carbon monoxide were produced on the other side, as a result of the oxidation of methane, indicating that hydrogen production by water splitting occurs because of oxygen transport across the membrane and not just surface reaction with the perovskite. This is confirmed by performing detailed carbon, hydrogen and oxygen material balances.

Acknowledgements

We wish to acknowledge the support of the EPSRC SUPERGEN programme through grant EP/G01244X/1 and the group of Professor Kang Li of Imperial College London for supplying the membranes.

REFORMER AND MEMBRANE MODULES (RMM) PRE-INDUSTRIAL PLANT FOR NATURAL GAS STEAM REFORMING

Dr. Marcello De Falco, Faculty of Engineering, University Campus Bio-Medico of Rome, Rome, IT
Dr. Gaetano Iaquaniello, Tecnimont KT S.p.A., Rome, IT
Dr. Annarita Salladini, Processi Innovativi Ltd, L'Aquila, IT

Abstract

A reformer and membrane modules (RMM) test plant with a hydrogen capacity of 20 Nm^3/h has been designed and built to investigate the performance of said innovative architecture at an industrial level. A major benefit of the proposed RMM configuration is the shift in the chemical equilibrium of the steam reforming reactions by removing the hydrogen produced at high temperatures, thanks to the integration of highly selective Pd-based membranes. By this way, the process can operate at a lower thermal level (below 650°C in comparison to the 850-950°C temperature needed in traditional plants).

Three types of Pd-based membranes are tested in order to compare performance in terms of permeated hydrogen flux. Moreover, a noble metal catalyst supported on a SiC foam catalyst is placed inside the reactor in order to improve thermal transport inside the reforming tubes.

The paper introduces the plant design criteria: the process scheme, the construction engineering of reformers and membrane units and the control system implemented to maximize experimental outputs. Then, some experimental tests results are reported and discussed. About 1000 operating hours and more than 70 heating and cooling cycles were performed. The average H_2 permeability for membranes tested are calculated and compared, and permeability expressions are reported. An overall feed conversion of 57.3% was achieved at 610°C, about 26% higher than what can be achieved in a conventional reformer at the same temperature, thanks to the integration of selective membranes.

The 20 Nm^3/h RMM installation makes it possible to completely understand the potential of selective membrane application in industrial high-temperature chemical processes, and represents a unique installation worldwide.

Keywords: hydrogen production, selective membrane, steam reforming, catalyst foam

* Corresponding author, m.defalco@unicampus.it

Introduction

Hydrogen is a key chemical building block in many chemical processes such as refining, ammonia and methanol synthesis. Methods to produce hydrogen from natural gas are well developed and account for over 48% of all hydrogen produced in the world. Steam reforming of natural gas is the most widely used process for efficient and economical benefits.

The reactions that occur during this process are as follows:

$CH_4 + H_2O \leftrightarrow CO + 3H_2$	Methane steam reforming	$\Delta H^\circ_{298K} = 206$ kJ/mol	(1)
$CO + H_2O \leftrightarrow CO_2 + H_2$	Water gas shift	$\Delta H^\circ_{298K} = -41$ kJ/mol	(2)
$CH_4 + 2H_2O \leftrightarrow CO_2 + 4H_2$	Overall reaction	$\Delta H^\circ_{298K} = 165$ kJ/mol	(3)

Steam reforming process is highly endothermic and equilibrium limited. To sustain global endothermic reaction and achieve high feed conversion, it is necessary to burn a part of methane feedstock in furnaces and operate at high temperature (800–850°C). This implies a reduced process global efficiency, increased greenhouse gas (GHG) emissions and a stronger dependence of hydrogen cost on the natural gas cost. A solution to overcome the equilibrium limitation is to selectively remove hydrogen from the reaction zone shifting reaction towards products. The integration of hydrogen selective membranes makes it possible to enhance hydrogen yield at lower temperatures reducing the thermal duty required to support the reaction. Pd based membranes and its alloys show excellent capacity of extracting hydrogen from a mixture of gases thanks to the high selectivity and permeation flow.

Several technique such as electroless plating, physical vapor deposition, spray pyrolysis and sputter deposition technique have been studied for the fabrication of Pd-based membranes. Most efforts have been focused on how to reduce the membrane thickness for maximizing hydrogen flux and reduce material cost while retaining mechanical integrity during high temperature cycles. Despite many attempt to develop thermally stable thin membrane, the maximum temperature is limited to 500°C (1).

Recent studies are focused on the development technologies for depositing a stable ultra-thin defect-free Pd based membrane on a substrates in order to assure a good mechanical support and maximize permeation flux (2-9). Among the last forty year pure hydrogen permeance has been increased of about an order of magnitude primarily due to a reduction of film thickness resulting from improved substrates and advanced fabrication techniques (9).

Membrane for hydrogen separation can be integrated in steam reforming process (SR) using two possible configurations: by assembling the selective membrane directly in the reaction environment (Membrane Reactor-MR) (10-13) or through a series of reaction and separation modules (Reformer and Membrane Module-RMM) (14-16). In a MR configuration, membrane thermal stability limits working temperature below 500°C while in the RMM configuration, operating temperatures in the reactor and separator are de-coupled and can be set separately (17).

A prototypal RMM pilot plant was developed and constructed in order to investigate the real performance of such innovative architecture. The plant, with a design capacity of 20 Nm^3/h, is based on two-step reformers, with the outlet reforming temperature working at 550-650°C, and two membrane modules working at 400-450°C.

Plant design criteria

Process plant description

The plant process scheme is illustrated in Figure 1 and consists of two-step reformer and membrane modules. Each reformer consists essentially of two main sections: the radiant box, containing the tube charged with the structured catalyst and the convection section, where heat is recovered from the flue gases for preheating feed and superheating the process steam. Design of the radiant chamber is quite conventional, it differs only by the heated length of the reformer tube which is around 3 meter, the tube metallurgy and the contained catalyst. One of the advantages of the proposed architecture is to require low cost stainless steel instead of exotic and quite expensive material as HP25/35 chromium/ nickel alloy.

Natural gas from battery limits or from cylinders at 20 barg is introduced through the pressure regulator and flow controller to the feed desulphurisation (DS) reactor, where sulphur compounds are removed. The desulphurised feed is then mixed with steam in ratio ranging from 3 to 4. Mixed feed is preheated in the convection section and enters the first reforming. The reformed gas product is cooled down through the first air cooler and fed to the first separation module at the desired temperature. A retentate, recycled to the second reformer stage and a mixture of H_2 plus sweeping steam, are produced. Stream coming from the second reformer stage is cooled down through the second air cooler and routed to the second separation module. H_2 from both modules are mixed together and sent to final cooling and condensate separation. Retentate from the second stage is sent to the flare. The pressure of both shell and permeate sides are controlled using a back pressure regulators. Both membrane modules are protected for excessive temperature using a pressure relief regulator installed on the income lines. All the vent points are connected to main vent system and routed to the flare.

Heat of reaction in both reforming steps is provided by two independent hot gas generators in order to set the reforming temperatures as required by the tests. The flue gases coming from the hot generator may reach a temperature greater than 800°C.

A dedicated hot oil boiler produces satured steam for steam reforming reaction and membrane sweeping. Sample connections together with pressure, temperature and flow measurement point are located at the inlet and outlet of reformer and membrane modules to completely characterize RMM performance. Gas composition was monitored on-line. CH_4, CO and CO_2 concentrations were measured by the online infra-red NDIR multiple analyser, ABB URA14, meanwhile H_2 was analyzed using the thermal conductivity detector ABB Caldas 17.

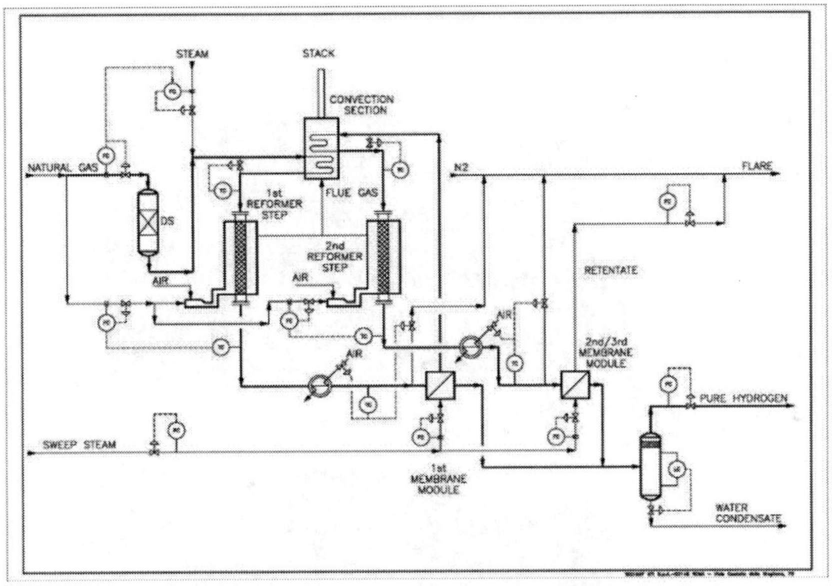

Figure 1. Process scheme of the test plant (18)

-Low temperature steam reforming catalyst

In order to enhance catalyst activity at low temperature, noble metals based on activated and stabilized Al_2O_3, supported on SiC foam, were selected for low temperature reforming instead of traditional nickel based. The catalyst is the SR-10, produced by BASF and composed by Rh-Pt. The open cross-flow also has the advantage of increasing effectiveness of the heat transfer through convection, reducing the tube metal temperature in a process where the rate-limiting factor is the heat transfer from the fluegas to the reforming tube wall. The use of a foam as catalytic support also results in a lower pressure drop through the tube which, in turn, reduces the compression energy cost of the feed.

SiC cylinder shaped foam, 150mm long and with a diameter of 60mm, is shown in Figure 2.

Figure 2. Steam reforming catalyst

Pd based membrane module

Three different Pd-based membrane separators were installed on the pilot plant. Among membrane producers, ECN from Netherland (19), MRT from Canada (20) and a company from Japan were selected and their membrane have been assembled in the steam reforming plant. Membranes have different supports and shape: ECN and Japanese manufacturer membranes are made by a ceramic support having a tubular geometry, while MRT membranes have a stainless steel substrate and planar geometry. All membranes are Pd/Ag alloy membranes with the exception of ECN, which is Pd made. Figures 3-5 show the installed modules.

Figure 3. Tubular ECN membrane separator (total installed surface 0.40 m^2)

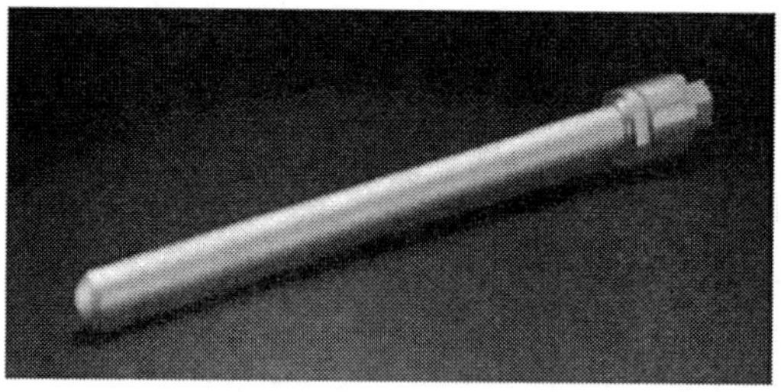

Figure 4. Tubular Japanese membrane separator (total installed surface 0.13 m^2)

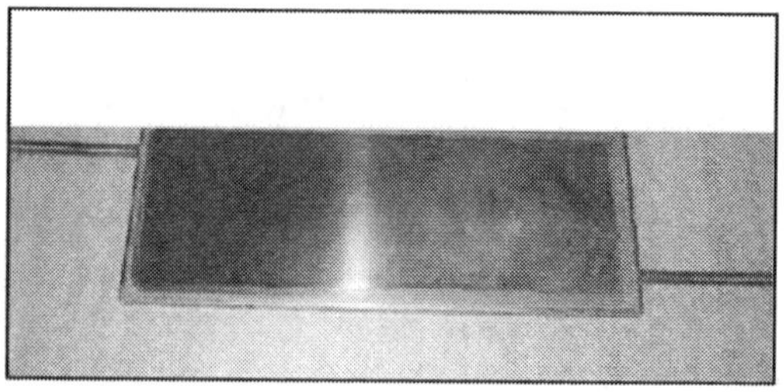

Figure 5. Planar MRT membrane separator (total installed surface 0.60 m^2)

Results and discussion

Low temperature steam reforming

Several test have been carried out on structured catalyst by varying process condition such as gas hourly space velocity (SV), reforming temperature (T_{ref}) and steam to carbon ratio (S/C). No sign of deactivation has been detected over 1000 hours of testing and more than 70 heating and cooling cycles. Figure 6 shows feed conversion and residue CH_4 along all over testing period at a reforming temperature ranging from 600 to 630°C, reforming pressure (P_{ref}) of 10 barg and steam to carbon ratio 4.8. Composition detected at the outlet of reforming reactor is always much close to the equilibrium.

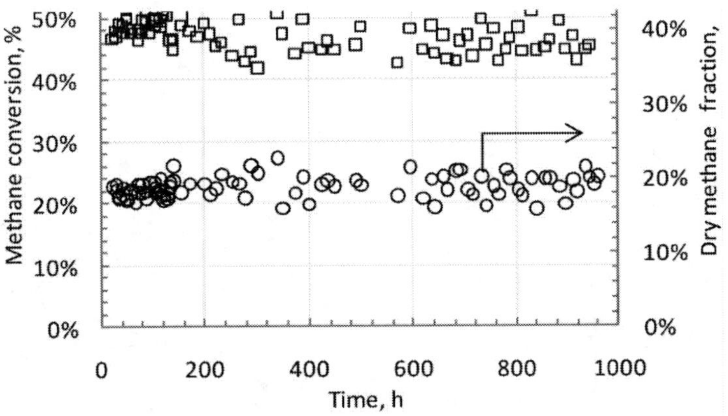

Figure 6. Feed conversion and residue CH_4 along testing period, T_{ref}=600-630°C, P_{ref}=10barg, S/C=4.8

Monitoring system allows to evaluate the influence of several process parameters. For example figure 7 shows effect of steam to carbon ratio over methane conversion at a reforming temperature of 600°C, reformer pressure of 10barg and space velocity of 5150h^{-1}. Higher S/C allows to achieve higher feed conversion. An increase of S/C from 3.8 to 4.8 allows to enhance methane conversion from 39% to 44.8%.

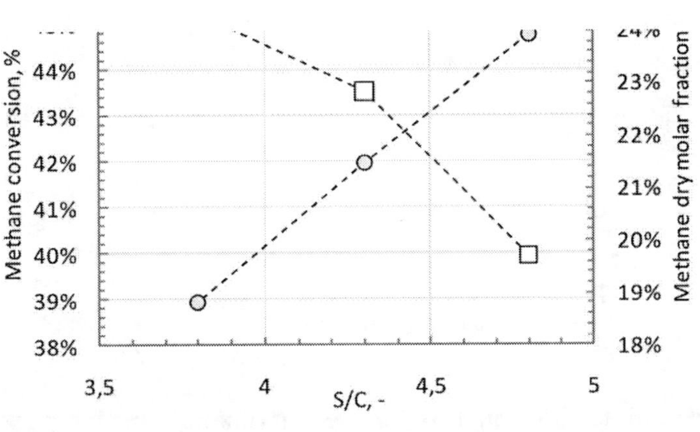

Figure 7. Feed conversion and residue CH_4 vs S/C, T_{ref}=600°C, P_{ref}=10barg, SV=5150h^{-1}

Pd based membrane module performance

In order to prevent swelling and damage, the membrane should not be heated or cooled in the presence of hydrogen when membrane surface temperatures are below 350°C. Heating up and cooling down were carried out under nitrogen flow with a slow heating rate (lower than 3°C/min) to avoid thermal stress.

At a fixed pressure difference between shell and lumen (10 barg in retentate, 0.4 barg in permeate zone), the hydrogen flux was measured for different operating temperatures within the range 400-450°C. The three membranes showed good stability under the real environment of the reformed gas mixture. As can be observed in Figure 8-10 no sign of flux decay was observed all over testing period. At a fixed pressure difference permeance is visibly influenced by membrane temperature.

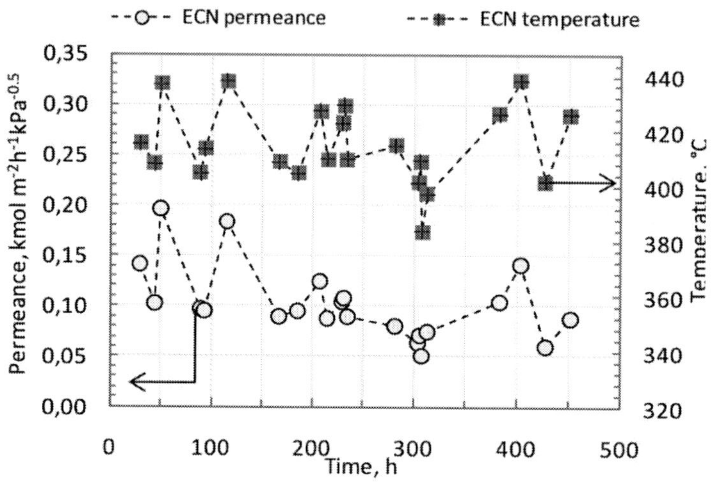

Figure 8. ECN membrane permeance vs time test and membrane temperature (Tm), T_{ref}=600-620°C, P_{ref}=10barg, S/C=4.8

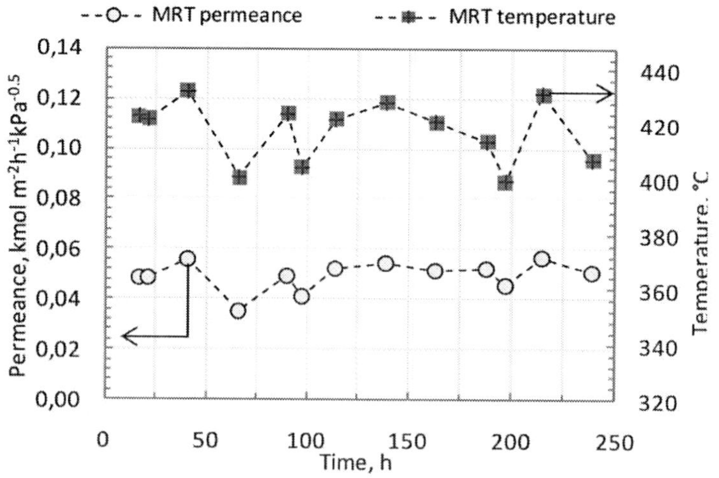

Figure 9. MRT membrane permeance vs time test and membrane temperature T_{ref}=600-620°C, P_{ref}=10barg, P_{perm}=0.4barg, S/C=4.8

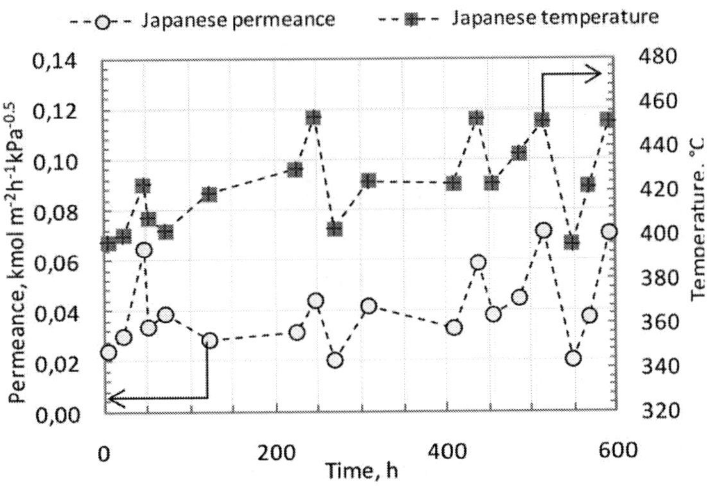

Figure 10. Japanese membrane permeance vs time test and membrane temperature, T_{ref}=600-620°C, P_{ref}=10barg, P_{perm}=0.4barg, S/C=4.8

Assuming the Sieverts' law and Arrhenius permeability dependence on the temperature, the hydrogen flux is given by the following expressions (4-5):

$$J_{H_2} = \frac{Perm}{\delta} \cdot \left(p_{H_2,feed}^{0.5} - p_{H_2,permeate}^{0.5} \right) \quad (4)$$

$$Perm = A \cdot \exp\left(\frac{-E_a}{RT}\right) \quad (5)$$

where J_{H2} is the hydrogen flux permeated, δ is the membrane thickness, $P_{H2,feed}$ and $P_{H2,permeate}$ are hydrogen partial pressures in the feed and permeate side. Activation energy and pre-exponential factor were reported in a previous work (18) and summarized in Table 1.

Table 1. Permeability for three installed Pd-based membrane

Membrane	A(kmol/mhkPa$^{0.5}$)	Ea (kJ/mol)	R^2
ECN	1.72E-1	77.0	0.705
MRT	5.75E-4	35.3	0.611
CONFIDENTIAL	9.31E-2	80.4	0.844

RMM architecture performance

Coupling two reforming stage and an intermediate membrane separation module allows to overcome equilibrium conversion. The RMM architecture applied to ECN membrane, performed a methane conversion 10-12% higher than equilibrium values as reported in Figure 11. An overall feed conversion of 57.3% was achieved at 610°C, about 26% higher than what can be achieved in a conventional reformer at the same temperature.

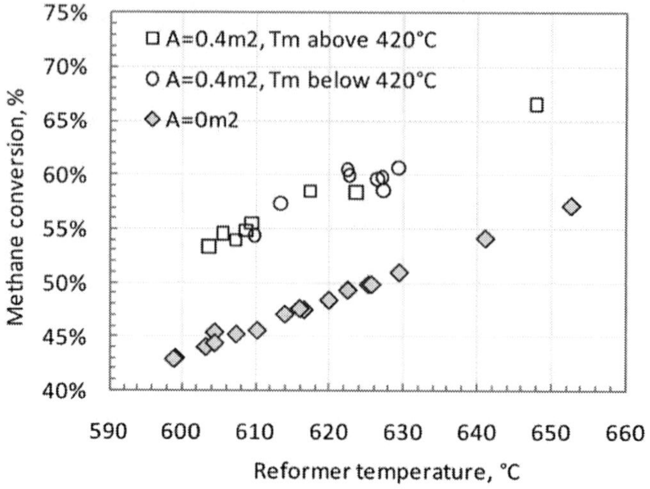

Figure 11. Methane conversion with and without membrane stage (ECN), P_{ref}=10barg, P_{perm}=0,4barg, S/C=4.8, A=0.4m^2

Figure 13 reports the effect of membrane thickness on methane conversion of the RMM architecture. A reduction of membrane thickness from 2.5 μm to 0.5 μm may enhance methane conversion of 10% due to the higher hydrogen removal. Thickness thinner than 0,5 μm have no more effect on global performance. The effect of membrane permeability on feed conversion is also reported in Figure 13 where experimental permeability is depicted as a=1. Obviously, at a fixed membrane thickness, an increase in permeability allows to obtain higher flux and consequently higher feed conversion. At a thickness of 2,5μm, membrane with a permeability higher than a factor of 1.5 and 3 may increase methane conversion till 4% and 8%.

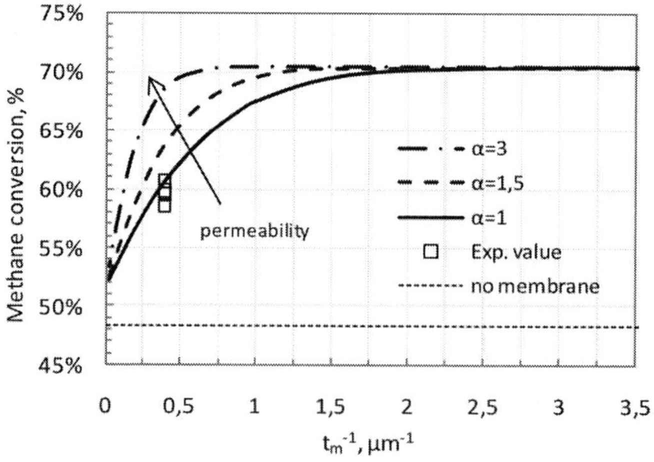

Figure 13. Effect of membrane thickness and membrane permeability on methane conversion P_{ref}=10barg, T_{ref}=630°C, S/C=4.8, A=0.4m^2

Another parameter allowing to realize high hydrogen flux is membrane temperature. Figure 14 shows effect of this parameter combined with membrane thickness. Methane conversion may be increased of about 10% rising membrane temperature from 420°C to 500°C. The same feed conversion obtained for ECN membrane at 420°C, may be realized with a thickness equal to 10μm at a 500°C.

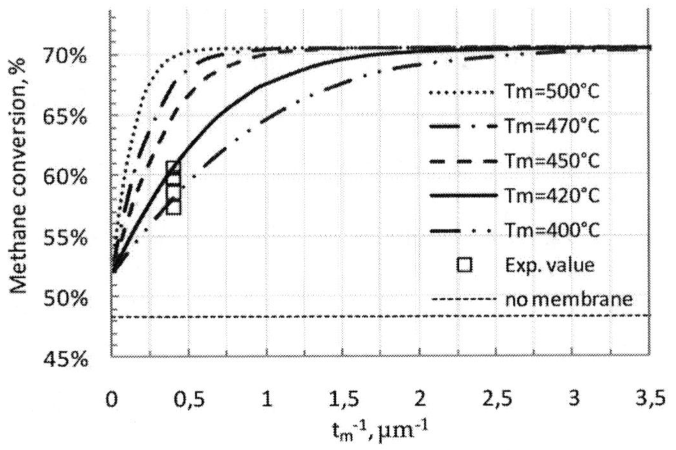

Figure 14. Effect of membrane temperature and thickness (t_m) on methane conversion, P_{ref}=10barg, Tref=630°C, A=0.4m^2, S/V=4.8, Tm=420°C.

Conclusion

An innovative architecture for steam reforming based on the concept of reformer and membrane modules has been developed and tested. Thin Pd membranes were tested in an industrial environment using the syn-gas produced by steam reforming of natural gas for more than 1000 hours and no less than 70 heating and cooling cycles. There was no significant sign of change throughout the whole test period for supported foam catalyst as well as for membrane modules. Experiments were successfully carried out in which an average enhancement of natural gas conversion over the conventional SR reactor ranging from 10 to 11 point percent was obtained. Higher feed conversion may be obtained using thinner membrane. A reduction of membrane thickness from 2.5 μm to 0.5 μm may further enhance methane conversion of 10% due to the higher hydrogen removal efficiency. A further increase of feed conversion equal to 15 and 19% may be obtained using membrane having a permeability higher than a factor of 1,5 and 3.

The good results of pilot plant demonstration indicates the promising feasibility in industrial application since the operating stability is also an important factor for the industrial application of the Pd-based membrane.

Acknowledgements

This work was carried out within the framework of the project entitled "Pure hydrogen from natural gas reforming up to total conversion obtained by integrating chemical reaction and membrane separation", funded by MIUR (FISR DM 17/12/2002)-Italy.

References

1. Okazaki K., T. Ikeda, D. A. Pacheco Tanaka, K. Sato, T. M. Suzuki, F. Mizukami (2011), "An investigation of thermal stability of thin palladium-silver alloy membranes for high temperature hydrogen separation", *J. of Membrane Science,* 366, 212-219.
2. McLeod L.S., F.L. Degertekin, A.G. Federov (2009), "Determination of the rate-limiting mechanism for permeation of hydrogen through microfabricated palladium-silver alloy membranes", *J. Membr. Sci.* 341, 225-232.

3. Okazaki J., T. Ikeda, D. A. Pacheco Tanaka, T. M. Suzuki, F. Mizukami (2009), "In situ high-temperature X-ray diffraction study of thin palladium/α-alumina composite membranes and their hydrogen permeation properies", *J. of Membrane Science*, 335, 126-132.
4. McCool B.A., Y.S. Lin (2001), "Nanostructured thin palladium-silver membranes: effect of grain size on gas permeation properties" *J. of Materials Science*, 36, 3221-3237.
5. Zang Y., J. Lu, R. Maeda, C. Nishimura (2007), "Microfabricated Pd and Pd-25Ag alloy membranes" *J. of Alloys and compounds*, 446-447, 579-582.
6. Klette H., R. Bredesen (2005) "Sputtering of very thin palladium-alloy hydrogen separation membranes" *Membrane Technology* 2005, 7-9.
7. Ke Z., G. Huiyuan, R. Zebao, L. Yeusheng, L. Youngdan (2007), "Preparation of thin palladium composite membranes and application to hydrogen/nitrogen separation", Chin. J. Chem. Eng. 15(5), 643-647.
8. Huang Y., R. Dittmeyer (2007), "Preparation of thin palladium membranes on a porous support with rough surface", *J. of Membrane Science*, 302, 160-170.
9. Hatlevik O., S.K. Gade, M.K. Keeling, P.M. Thoen, A. P. Davidson, J. Douglas Way (2010), "Palladium and palladium alloy membranes for hydrogen separation and production: history, fabrication strategies, and current performance", *Sep. Purif. Technol.*, 73, 59-64.
10. De Falco M., "Pd-based membrane steam reformers: A simulation study of reactor performance", *Int. J. of hydrogen energy* 33 (2008) 3036-3040.
11. Fernandes Fabiano A. N., A. B. Soares, "Methane steam reforming modelling in a palladium membrane reactor", *Fuel* 85 (2006) 569–573.
12. Gallucci F., L. Paturzo, A. Basile, "A simulation study of the steam reforming of methane in a dense tubular membrane reactor", *Int. J. of Hydrogen Energy* 29 (2004) 611–617.
13. Lin Y., S. Liu, C. Chuang, Y. Chu, Effect of incipient removal of hydrogen through palladium membrane on the conversion of methane steam reforming: experimental and modelling, Catal. Today 82 (2003) 127-139.
14. De Falco M., D. Barba, S. Cosenza, G. Iaquaniello, A. Farace, F. G. Giacobbe, Reformer and membrane modules plant to optimize natural gas conversion to hydrogen, Special Issue of Asia-Pacific J. Chem. Eng. on Membrane Reactors (2009), DOI:10.1002/apj.241.
15 De Falco M., D. Barba, S. Cosenza, G. Iaquaniello, L. Marrelli, Reformer and membrane modules plant powered by a nuclear reactor or by a solar heated molten salts: assessment of the design variables and production cost evaluation, *Int. J. of Hydrogen Energy* 33 (2008) 5326-5334.
16. Li A., C. J. Lim, J. R. Grace, Staged-separation membrane reactor for steam methane reforming, Chem. Eng. J. 138 (2008) 452-459.
17. De Falco M., G. Iaquaniello, B. Cucchiella, L. Marrelli, Syngas: Production Methods, Post Treatment and Economics. Reformer and membrane modules plant to optimize natural gas conversion to hydrogen, Nova Science Publishers 2009, ISBN: 978-1-60741-841-2.
18. De Falco M., G. Iaquaniello, A. Salladini (2011), "Experimental tests on steam reforming of natural gas in a reformer and membrane modules (RMM) plant" *J. of Membrane Science*, 368, 264-274.
19. www.hysep.com, Hysep 1308 module.
20 www.membranereactor.com

(110a) Why Does It Cost So Much? Technical and Non-Technical Factors That Affect Early Stage Cost-of-Manufacturing Estimates of Developing Processes

John Eric Anderson, Core R+D, Engineering and Process Sciences, Dow Chemical, Midland, MI and Thomas I. Gilbert, Core R+D, Dow Chemical, Midland, MI

Why Does it Cost So Much? Technical and Non-Technical Factors that Affect Early Stage Cost-of-Manufacturing Estimates of Developing Processes

John Anderson (jeanderson@dow.com)

and

Thomas Gilbert (tigilbert@dow.com)

Dow Chemical

Abstract

Estimates of the manufacturing cost of new products and processes at very early stages can be very useful when decisions are made to support or abandon projects. If historical cost performance data from the estimator's company or organization is available, it can be used as the basis of an extrapolation to predict the cost of a process being developed. This process does require some 'normalizing' of the cost data so that a valid extrapolation is made. Volume, process complexity, location and other factors should be considered and the impact of these variables will be reviewed. In those cases where the production is to be executed at a third party, the extrapolation process also requires some insight about the outside firm's technical capabilities and business practices. For many reasons, prediction of cost and pricing at third parties, while very useful, is particularly challenging. Qualitative observations made about the additional sources of variability observed in these situations and attempts made to quantify these will be discussed.

(110b) Energy Consumption Analysis for An Ethylene Plant Start-up

Chaowei Liu, Xiongtao Yang, Qiang Xu and Kuyen Li, Dan F. Smith Department of Chemical Engineering, Lamar University, Beaumont, TX

Chemical process energy systems consist of the equipment necessary for process heating, process cooling and refrigeration, electro-chemical processes, mechanical operations, and other direct process uses. At the same time, energy losses occur during the normal process operations and also the turn around operations. Many studies have been conducted on the energy distribution and energy efficiency improvement for normal chemical process production. However, quantitative studies of the energy consumption analysis on chemical plant start-ups are still lacking.

Through the plant-wide dynamic simulation, this paper studies the quantitative energy performance for an ethylene plant start-up. Details about the energy profile of cooling and heating duty of distillation towers, auxiliary heat exchanger duty for start-up, power consumption in charge gas compressor system are provided with respect to the start-up operation procedures. This study enhances the energy research for the manufacturing energy studies and it will also help analysts and decision-makers to better understand the energy consumption in chemical plant turn around operations.

(110c) Pilot Plant Retooling for a More Reliable Process Scale-up

Pankaj S. Gautam, Process Technology, SABIC Innovative Plastics, Mt. Vernon, IN

Piloting a new process at a smaller scale before rolling it out to the larger manufacturing facility presents numerous scale-down as well as scale-up challenges. This paper will focus on several case studies to illustrate best practices we learned through experience for a successful scale-up. The paper will cover the following case studies –

1. Improving liquid-liquid separation through the use of a coalescer centrifuge and scale-up of the same
2. Lowering the contamination of polymer with residual monomer through modifications in reactor design.
3. Enhancing pilot plant capability through the use of a concentrator

A Treatise of Filter Cake Washing Mechanisms In Pressure and Vacuum Filtration Systems

Barry A. Perlmutter, President & Managing Director
BHS-Filtration Inc.
9123-115 Monroe Road
Charlotte, North Carolina 28270
Phone: 704.845.1190
Fax: 704.845.1902
E-mail: barry.perlmutter@bhs-filtration.com

Introduction

During many chemical and pharmaceutical process operations, cake washing is required since no filter system can remove all of the mother liquor by pressure or vacuum dewatering only. The need for cake washing to produce a "clean cake" for further processing, drying or disposal can be for several reasons including the removal of the mother liquor, exchange of solvents, dissolution of soluble components and the displacement of insoluble contaminants.

The topic of "how to most efficiently wash the cake" is one that many process engineers struggle with when considering the various techniques available to meet the economic, production and quality requirements of the operation. Chemical engineering textbooks are notoriously shy on the subject and such information that is available is frequently lost in a host of theoretical equations. This article introduces pragmatism into cake washing theory and will help to guide engineers during their decision-making processes.

Selection of Equipment

Before discussing cake washing, it is necessary to examine some of the equipment that can produce a cake, from thin-cakes (from 3 mm – 25 mm) to thick cakes (up to 300 mm). It can be suggested that any filter can produce a cake to be washed. Given an infinite amount of time and patience, yes perhaps this is true; the real question, however, is "does the process have enough time and do the operators have enough patience?"

The obvious choices for cake-producing filters that are also capable of washing the cake are well documented in the literature. These include rotary pan filters, "box" filters, filter presses, belt filters, centrifuges, rotary drum filters and nutsche filters. Whether the application is, in principle, better suited for one rather than another depends upon many factors including the process requirements, cost, space availability, etc. The equipment selection should occur based upon the overall balance of the objectives and should include bench top and pilot scale testing.

However, to evaluate if the selected filter will meet the cake washing requirements, there is no substitute for painstaking laboratory tests simulating, in the best possible manner, the conditions of the pre-selected filter and then, based upon the washing results, decide if this filter remains the appropriate choice. If not, then the filter selection must be re-evaluated.

Cake Preparation for Optimum Washing

It is a logical assumption, but in most cases erroneous, that the drier the cake the easier the cake can be washed. The reason for this apparent contradiction is either one or both of the following:

- If the cake is dewatered to its maximum through compression, the cake almost certainly will be so compact that no wash liquid can penetrate it. The wash liquid will look for the path of least resistance which means it will simply by-pass the cake or find the slightest crack and disappear through that crack.
- If the cake is dewatered beyond the dry-top by vacuum or blowing pressure gas (or steam) through it, the result will be the inevitable pockets (or cracks), which if anything will, once again, provoke bypassing.

In general, cake washing should begin immediately after the mother liquor has been drained away, leaving a well-settled, well-formed filter cake, which is saturated with mother liquor. If the surface of the cake (dry-top) shows signs of drying out, then it is usually too late.

Washing Mechanisms: Displacement and Dilution

In almost all cases, the requirement for maximum wash efficiency is to develop a mechanism, which removes the required amount of mother liquor with a minimum of washing liquor. Basically there are only two mechanisms, and in practice usually both are present:

- Displacement of the mother liquor by washing liquor
- Dilution of the mother liquor through mixing with washing liquors

Displacement is by far the most efficient mechanism and under ideal conditions would cause the washing liquor to act as a piston driving the mother liquor out of the cake without any dilution. This is clearly utopia although a high degree of "piston or plug-flow effect" can be achieved with a careful preparation of the cake and a careful application of the wash liquor. The trick lies above all in the careful application of the washing liquor to the saturated face of the filter cake.

In open filters like belt filters or rotary drum filters, one can visually observe how far the wash application disturbs the cake surface and thus prevents good displacement. In closed filters this is more difficult and a degree of guesswork – and checking the result of these guesses – is inevitable.

If, as is very common, the washing liquor has a lower viscosity than the mother liquor one must be careful not too apply too much pressure (positive or negative) because the (usually) lower viscosity washing liquid will find any slight crack or "easy" passage between particles in preference to pushing the mother liquor.

Finally, cake washing will practically always alter the structure of the cake (change of polarity, change of viscosity, change in temperature, physical disturbance with the introduction of the washing liquor, etc.). As a result there comes a point where displacement washing all but stops and a dilution process begins.

Dilution can occur in three different scenarios:

- If the liquors are readily miscible and the cake liquor (by now a mixture of mother liquor and wash liquor) is not too much occluded inside the particles, its crevices or its clusters, it should be a fairly straightforward dilution process with a fairly predictable curve.

- If the liquors are reasonably miscible and the mother liquor has no great affinity to the solid particles, but is held inside the particle crevices or particle clusters, then a degree of agitation can help to "tumble" the cake a little and scrub the residual mother liquor off the particles. Obviously this agitation should only start after the "free" cake liquor has been washed away and when one is left with only the occluded liquor held in the pores of the particles or between particles. This agitation action is, of course, very difficult to imagine, for example, in filter presses or centrifuges. Two exceptions to this would be a multi stage pusher centrifuges that "tumble" the cake as matter of fact or a pressure plate filter with vibrating plates rather than spinning plates.

- If the mother liquor has a great affinity to the solids and is held inside the pores of the solid particles, usually the easiest and most effective washing system is to remove the cake from the filter, re-slurry it with clean washing liquid and allow it time to effect the desired mixing / diffusion and then to re-filter the slurry.

An Alternative Final Washing Scheme: Cake Compression versus Dilution

Since dilution is proportional to the volume ratios, the efficiency in terms of residual mother liquor being removed drops dramatically as the cake becomes cleaner (less mother liquor). Assuming a wash volume equal to the cake liquor volume, one can expect at best a 50% reduction. Initially this may be quite high, but by the time the cake only holds 0.1% of mother liquor, a subsequent wash would only remove 0.05%. With the present trend to produce purer end products, there may come a point where extremely large amounts of wash quantities and time would be required to dilute the last traces to meet such purity levels. In these cases, there is great merit in investigating compression of this (almost pure) cake, with or without gas blowing, and mechanically forcing part of the remaining diluted liquor out of the cake.

Applying Washing Liquors and Pitfalls to Avoid

The objective for washing is to subject the entire cake face to an even and evenly pressurized liquid wash. Certain filters offer better flexibility than others to achieve this effect. Above all one has to avoid "channelling" both through the cake and across its surface. (Dye tests will show this clearly – see "laboratory tests section"). If channelling does occur, it may be necessary to investigate a different type of filter.

It must be clear, but is often ignored, that washing at too high a pressure differential only leads to poor efficiency. Too high a pressure almost always leads to bypassing. Although the ideal cake has an even structure, no pinholes, and no cracks and lies firmly bedded against the retaining walls, we are not living in an ideal world and to some degree all cakes will have these faults. Forcing liquid through the cake at unnecessarily high pressures only makes matters worse. A pinhole will become a hole, a crack a ravine and if the cake does not bed down perfectly, the washing liquid will make a highway of the gap. In addition, high pressure differentials reduce the contact time. In all cases therefore the optimum pressure differential is the lowest possible.

With (continuous-indexing) horizontal vacuum filters, one can use overflow weirs to gently pour the liquid onto the cake to avoid cake disruption.

On all filters, after an initial displacement wash, the subsequent diffusion / dilution washes may benefit from misting sprays which often give a better and/or even "wetting" effect. The problem here is, however, that fine sprays may have a tendency to block the cake structure, whereas the coarser sprays may tend to disturb the cake.

In this respect, engineers may custom-design a highly effective arrangement when trying to optimize a washing technique. One such customer designed a unique spray system for a continuous-indexing vacuum belt filter. The system used a square pattern spray, angled at approximately 45º and placed it so that the jets lost their velocity at just the point where washing should take place and where thus a gentle rain of washing liquid dropped onto the cake, as shown in Figure 1. This is by far both a very inexpensive and efficient / gentle wash liquid applicator system and can be used equally on rotary drum filters, belt filters, and rotary pan filters etc.

Co-Current Washing versus Counter-Current Washing

As with filter selection, the decision for co-current or counter-current washing sometimes is based upon the history of the process. However, by careful analysis, the decision of co or counter-current washing can be made based upon strict objectives.

The co-current system is clearly simpler to install and to operate. It is also possible that with some special attention to details an acceptable overall efficiency can be obtained. In addition the co-current washing can be carried out on most filters whereas with counter-current washing one has to use filters of a large linear configuration, which in most cases means horizontal vacuum belt filters. Continuous – indexing designs offer the maximum separation of filtrates and hence the maximum efficiency for counter-current washing.

A well-designed counter-current washing scheme for a filter can achieve remarkable savings in wash liquor usage along with the corresponding economical savings. Since the reason for contemplating counter-current washing is practically always an economical one, a very careful balance has to be made for the total capital investment and running cost of both a co-current and counter-current system. Almost inevitably the counter-current system will be more costly to install, will take up more space and will have a higher energy cost; whether or not this is justified by the process gains, can only be decided by the process application.

Installing a Counter-Current Cake Washing Loop

The purpose of counter-current cake washing is to reduce as much as possible the quantity of washing liquid used for a given result. The reasons are practically always economical and for this reason it is even more important that the laboratory washing tests are being conducted under controlled and verified conditions. In most cases counter current washing is used on continuous equipment although it is perfectly feasible to do so on a number of batch filters, the principles remaining the same. A typical counter current washing train is shown in Figure 2, in this case using a continuous – indexing horizontal vacuum belt filter, which is probably the most commonly used filter type for good and controllable cake washing.

With the cake travelling from the left to the right, the clean washing liquid is introduced at the extreme right hand side, is collected and via pumps re-introduced as a washing liquid now containing some of the mother liquor onto the stage immediately before the last one etc. The increasing line thickness indicates the increase in concentration in the washing liquors, whereas the crosshatched "cake liquor" is shown tapering towards the end of the filter, indicating the removal of mother liquor.

At the Feed End of the filter, the two liquors, the concentrated washing liquid and the mother liquor, are being fed forwards either combined or as individual streams. This is the classical scheme which is shown in just about every textbook, but which ignores a very important alternative. If, as is common, a degree of displacement washing can take place after initial dewatering of the cake, it would not be terribly logical to do so with a washing liquid, which already contains a high concentration of mother liquor. The resultant cake simply would not be very clean.

In Figure 3, there is shown the alternative approach, which in most cases gives much better efficiencies. In this case (an arbitrary) 90% of the available volume of washing liquor is fed to the tail end of the filter as in the previous case. However, the remaining 10% is introduced as a displacement wash immediately after dewatering. This obviously leaves a much "purer" cake for the counter current washing stream, and thus can lead to either fewer steps or less washing fluid volume or both. It also eliminates one recycle loop complete with pumps, flow control etc.

In this respect it is important to remember that for the filter size the only thing that matters is the TOTAL hydraulic load on the filter and not the individual volumes, which are being recycled. If not watched carefully this can lead to excessively large filters.

For example, if in a co-current washing scheme, a cake can be washed with washing liquid equal to five (5) times the volume of the Mother liquor in the cake, then the same cake can be washed with only 1.5 volumes in counter current fashion; but while the counter current loop consists of five (5) stages, the total hydraulic load has increased by only 50% (1.5 x 5 = 7.5 times, as opposed to 5 times). In addition each loop consists, external to the filter itself, of a receiver, a pump, level control, flow control and the pipe work to go with it. All of this can be a sizeable percentage of the total capital investment for the project as well as of the operating costs.

One important point is that it is essential that the counter current stream be introduced to the filter cake evenly and in proportion to the volume introduced in the previous stage. This can lead to the installation of quite complex and expensive flow control.

To simplify this problem and reduce it to basic mechanics, a recycle loop, as shown in Figure 4, can be installed. It consists of a filtrate receiver, which in the case of vacuum filters is connected to a vacuum system or can be vented in case the filtrate drains from a pressure filter. A filtrate pump with an approximate 10% overcapacity runs continuously, delivering the filtrate to a Tee which connects with a float valve in the receiver and to a non-return valve and piping leading to the next washing stage. As soon as the pump overtakes the incoming filtrate, the float valve opens and the filtrate recycles preferentially back to the receiver, until the float valve closes and the filtrate goes to the next washing stage. In practice the float valve finds its own level resulting in an almost 100% constant flow of filtrate to the next stage. It is recommended that the riser above the non-return valve is 1 meter or more. If pipe work has been oversized then it may require a reducing valve above the non-return valve. This system has been used for more than 25 years in numerous installations and has always performed well, in spite of a complete absence of sophisticated control equipment.

Due to the fact that counter-current loops require a relatively large amount of space if individual recycling receivers are used, there is merit in using a single, multi-compartmented receiver as shown in Figure 5. Provided that all stages can operate at the same pressure differential, this arrangement is a design that can be used to minimize space and cost.

Laboratory Tests: Test Apparatus

How well a cake can be washed, if at all, depends 100% on the cake structure. Numerous articles have been written about cake construction or cake build-up. The theory is generally correct but in most cases only of academic interest, as the most important question that interests a process engineers is "our" cake and "our" process. What the process engineer must know is: "how well does this cake release its mother liquor under the influence of washing liquors". Of course, much more important than theories about the cake, is the use of the correct test apparatus, which must be able to simulate the actual working of the pre-selected filter.

For example, if one thinks of a rotary drum filter, one has to use a filter leaf, which can be dipped upside down in the suspension while for a horizontal vacuum filter one can use a standard filter funnel. Continuing these thoughts, for a centrifuge one has to use a laboratory centrifugal filter and for a filter press of whatever type one has to use a filter chamber into which the suspension is pumped at the required pressure. All these types will produce cakes of different structures and thus have different behaviours.

Dewatering / Cake formation

In the same way that an actual filtration curve is impossible to calculate without having carried out basic filtration tests on the appropriate test equipment, cake washing is equally impossible to calculate without laboratory data. Any filtration and or washing tests that are conducted must produce filter cakes of the thickness, which one expects under actual plant conditions. If one expects to use a filter press producing 30 mm cakes but carries out washing tests on a 15 mm thick cake, one cannot be sure of the results.

As for the test apparatus, the size of the lab filter units can range from 50 mm (2 inches) up to 150 mm (6 inches) or even larger if space and suspension volumes permit. The edge effects, in small lab filters, can become a disproportionate factor giving highly misleading results. In addition if the cake has a tendency to crack, usually the result of shrinkage, this may not show up in a small unit, since there may not be sufficient mass to produce the shrinkage crack. It is critical that the engineer conducting the bench-top tests be an experienced filtration engineer from the specific vendor or from the operating company. Filtration experience, or the art of filtration, will have an influence upon the testing results.

For equipment with a given and reproducible cake thickness, the first stage of cake formation is fairly simple as there are not many possible variations; although the basic rule of all pressure filtration applies: Do not start immediately with a high pressure differential but build the pressure up slowly, allowing the cake to form evenly instead of being forced.

For equipment, which allows for different varying cake thicknesses, tests should be carried out at different thicknesses. The problem is that often the optimum cake for basic dewatering is not ideal for washing and vice versa. The reason for this is that a cake that drains easily and rapidly probably also passes the wash liquors rapidly without doing much mother liquor displacement. Conversely, an ideal "cake wash thickness" may cause the overall filtration and washing time to be too long for practical purposes; some compromises, in the art of filtration, are often inevitable.

Co-Current Washing Tests

After having established the basic filtration (dewatering) parameters, one can begin the painstaking process of introducing a given amount of wash liquor, draining the cake as far as possible, opening the filter and measuring the residual mother liquor in the cake.

One important factor is the procedure for introducing the wash liquor. If one uses a normal filter funnel (to simulate horizontal filters), then care must be taken to ensure that the washing liquor does not disturb the cake and its surface. At times this can be quite difficult and may require pouring the wash liquor along a spatula. An atomizing spray clearly helps greatly but it is more difficult to check the exact volumes which one sprays. An open (about 100 micron) monofilament cloth laid on top of the cake may protect the surface from being disturbed.

Measuring the mother liquor content in the filtrate is interesting but, in reality, is not a measure of cake washing efficiency. For example, a filtrate with a low level of mother liquor can indicate that the cake is washed or that most wash liquor by-passed the cake and did not do any washing at all.

In terms of the test pressures, the tests should be repeated for, at a minimum of at least three different pressure differentials. Almost certainly one pressure differential gives better washing than another. If this indicates a trend, follow that trend to the optimum condition. Thereafter different volumes need to be tested and ultimately one can build up a washing profile, which then can be the basis for the actual plant conditions.

Because mother liquor determination can be time consuming, and above all, to get a better understanding of the wash pattern, it is a great help to introduce a soluble dye in the suspension (fountain pen ink being often the easiest). This gives an immediate visual indication of the displacement wash effect (the wash filtrate is dark blue, although the wash liquor is colorless) and the effectiveness of the subsequent dilution steps. When taking the cake out of the filter (without disturbing the cake if at all possible) one can also see clearly how effective the washing has been. If the cake shows areas of dark blue then clearly there has been either a distribution problem with the wash liquor or in these areas the cake was too compressed, preventing penetration with wash liquor. Equally white "veins" indicate that there were, perhaps invisible, cracks through which the wash liquor travelled and washed that part and not much else.

Counter-Current Washing Tests

Before starting on a counter-current washing test, some co-current tests should be carried out to get a "feel" for the cake's behavior and the way the wash liquor removes the mother liquor, as well as for an approximate guess at the volumes required to obtain the desired results.

The principle of the counter-current washing laboratory tests is shown in Figure 6. As shown, three filter funnels are set up in one line. All funnels must have identical dimensions, have the same drainage grids, same filter cloth and be connected to the same vacuum source. They are marked A, B and C. To prevent confusion it helps to mark both the funnel and the flask. Sufficient suspension for four cakes must be available and be kept gently agitated and/or heated in case of hot suspensions. The initial order is B, C and A. Following the instructions one finally obtains representative washed cakes and filtrates. Representative, that is, for a three-stage counter-current washing scheme.

It will be noted that this test does not allow for a 1st displacement wash with clean wash fluid. It is rather complex to introduce this 1st step in a laboratory set-up and since the efficiency of a first displacement wash can be measured exactly, it is relatively easy to allow for this in the final calculations, once the efficiency of the whole counter current train has been established.

Clearly one can expand the test to include as many stages as one likes although in general a three stage counter-current wash is fairly optimal. A schematic for 4 stages is shown in Figure 7 and can, of course, be further extended to 5 – 6 or more stages. During all of these tests all filtrate volumes; wash volumes, wash times and cake purity must be recorded. Finally, it may not be obvious from these schematics, but these tests are always more time consuming than they first appear and therefore it is generally recommend that two trained operating engineers or the vendor and the process engineer perform the test work together.

Conclusion

Filtration and cake washing are affected by many variables. For example, in terms of the particles, the amount of solids, size, shape, particle density, compressibility of the solids, zeta potential and other ionic forces, agglomeration of the particles due to internal bonding and forces, etc. all impact the filtration flux rate and washing rate. Two particles of the same size, may behave differently if one is flat in shape while the other is irregularly shaped. In terms of the liquid, there are also characteristics that impact the filtration and washing rates such as temperature, viscosity, density, pH, components of the liquid, chemical additions (flocculants, coagulants, etc.), polar/non-polar constituents, the interaction of the solids and liquids, etc. Once again, a small change in the liquid, can have a dramatic impact on the rates.

As can be seen, many variables must be examined during filtration and washing studies. This article discussed techniques to conduct washing laboratory and pilot studies and then how these results can be transferred to production systems. The process engineer through the development of an optimum test plan can analyze the resultant cakes and slurries to maximize his or her process. If the actual results differ from the tested results, then, with a good baseline of tests, the engineer can analyze the differences and develop the necessary corrective actions for a successful result and process.

FIGURES

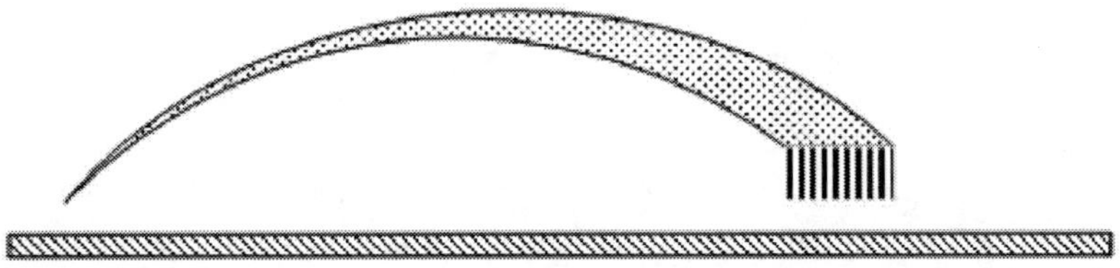

Figure 1: Custom-design and highly effective washing arrangement

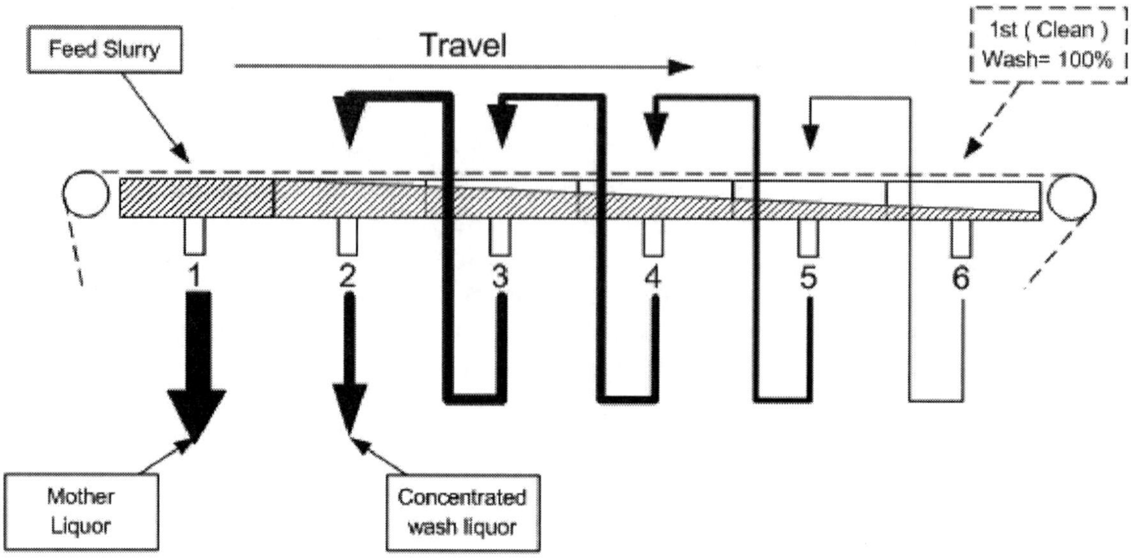

Figure 2: A typical counter current washing train

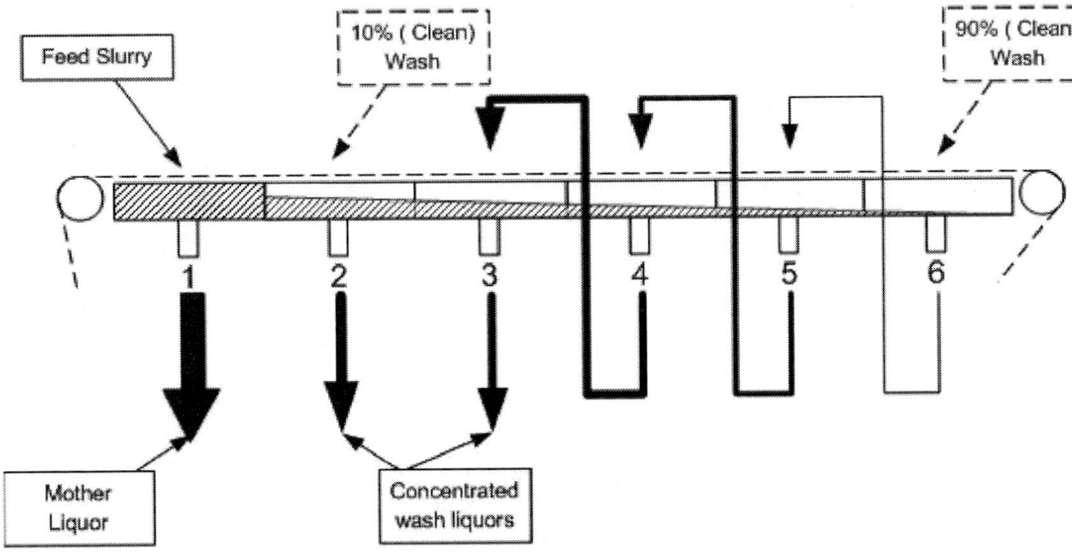

Figure 3: 90% of the available volume of washing liquor is fed to the tail end of the filter as in Figure 2. However, the remaining 10% is introduced as a displacement wash immediately after dewatering.

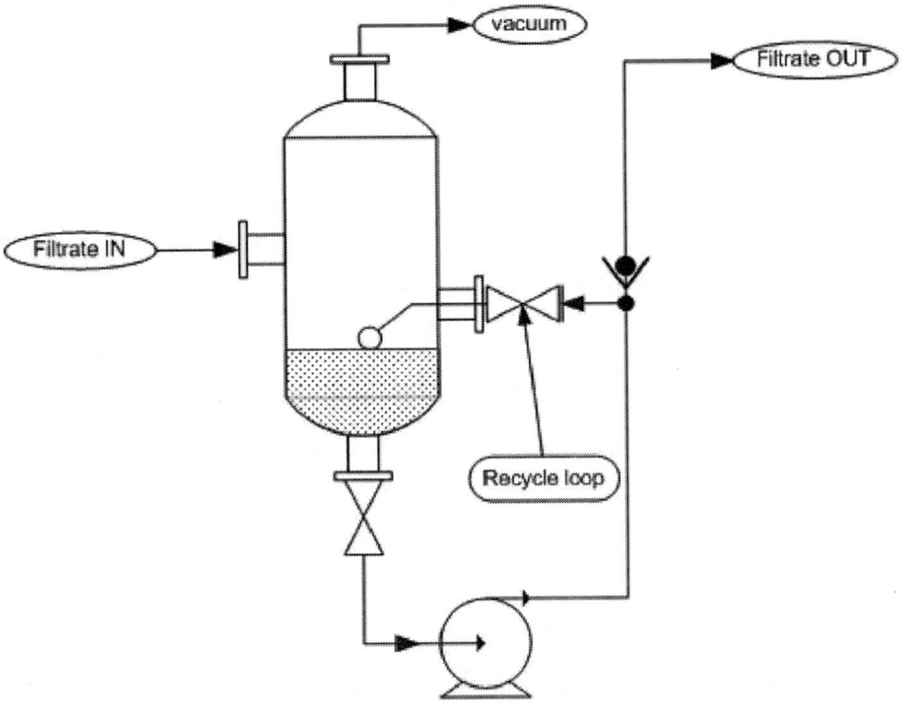

Figure 4: Typical Wash Liquid Recycle Loop

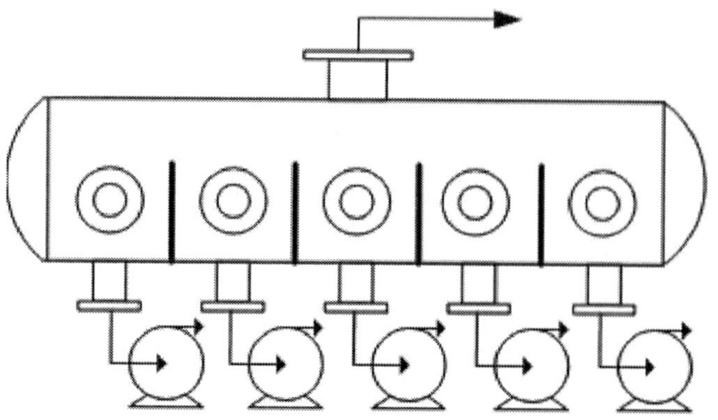

Figure 5: Typical multi-compartmented receiver as an alternative to individual receivers shown in Figure 4

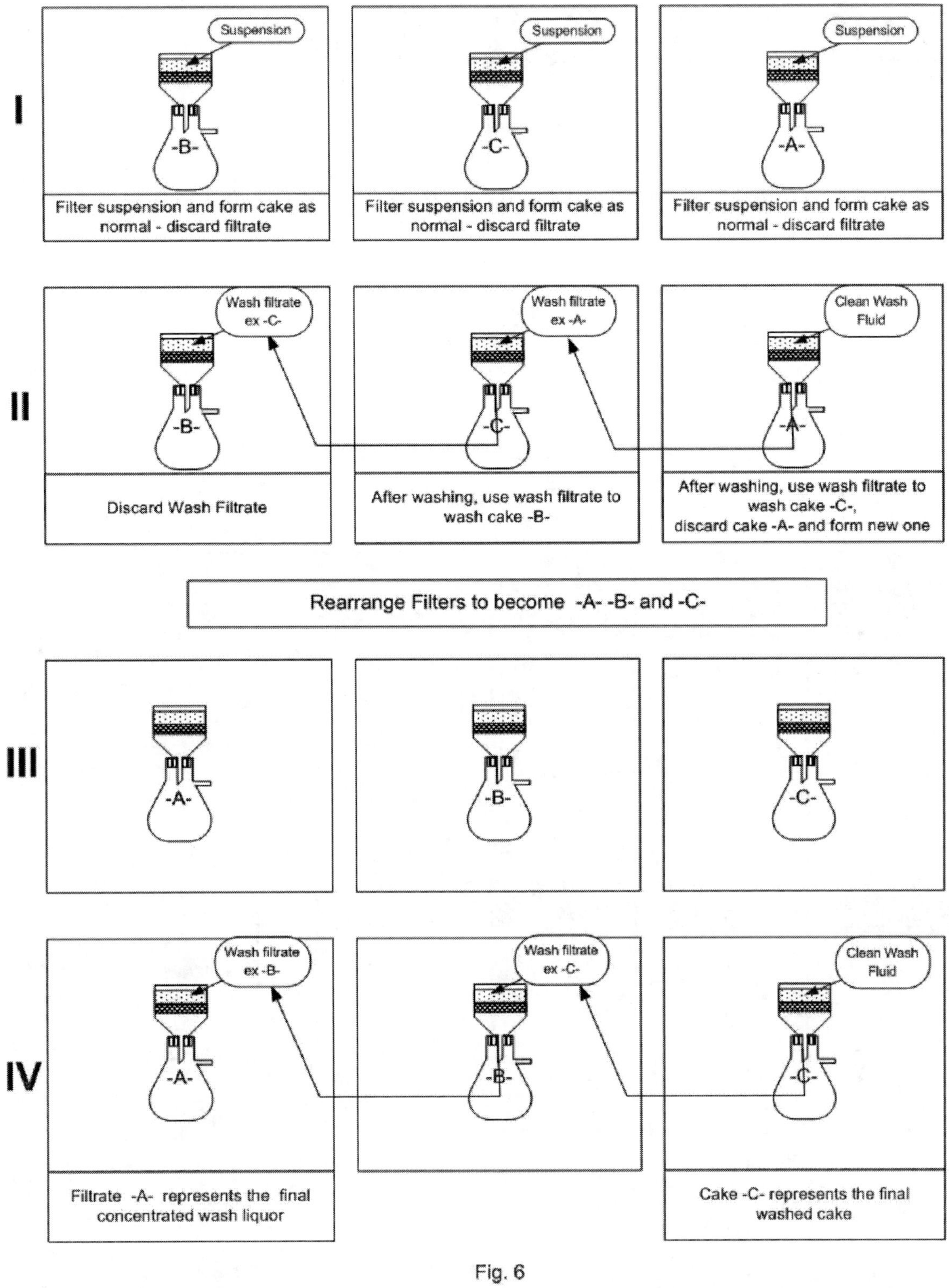

Figure 6: The principle of the counter-current washing laboratory tests

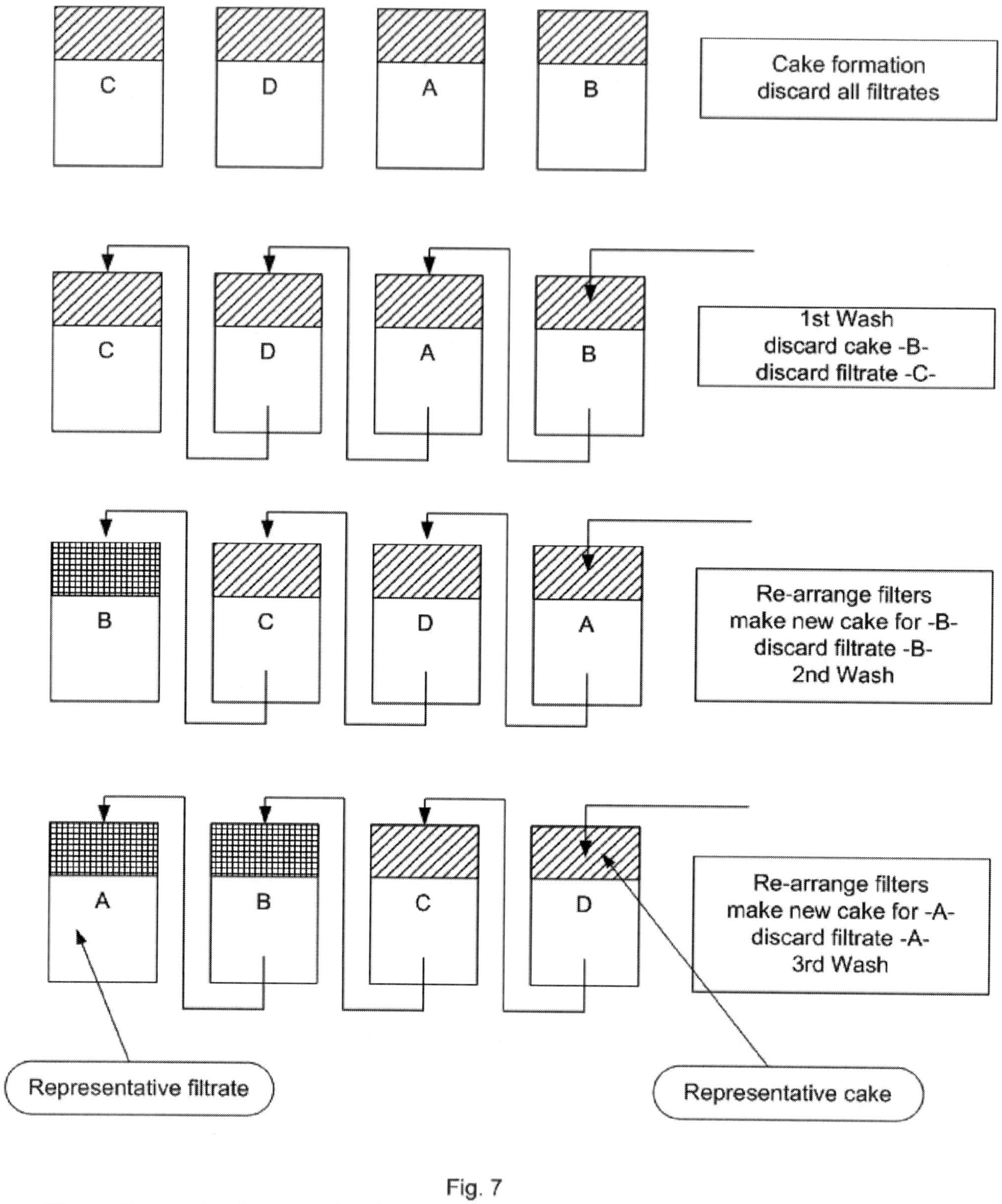

Figure 7: **A schematic for 4 stages of washing which can, of course, be further extended to 5 – 6 or more stages.**

Continuous Biodiesel Production from Animal Fats and Waste Cooking Oil Using Supercritical Methanol and Nanostructured Heterogeneous Catalysts

Maoqi Feng, Chee Kai Tan, Chemical Engineering Department, Chemistry and Chemical Engineering Division, Southwest Research Institute, San Antonio, Texas 78238, email: mfeng@swri.org

INTRODUCTION

As the availability of petroleum-derived diesel fuel slowly decreases, the interest in alternative fuels grows. Biodiesel refers to a vegetable oil or animal fat based fuel consisting of long-chain fatty acid alkyl (methyl, ethyl or propyl) esters. It is a renewable alternative fuel and has been successfully derived from lipids (vegetable oil or animal fat) and alcohol as shown in Equation 1.

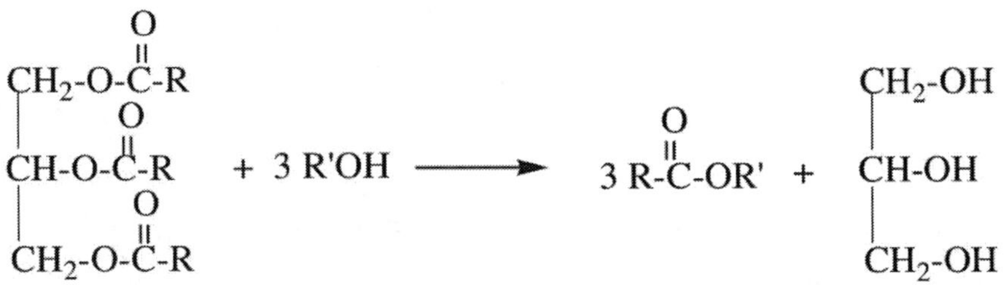

(1)

R = fatty acid chain

R' = CH_3

Biodiesel is becoming well-known as an environmentally friendly fuel due to its nontoxic and biodegradable characteristics. It has reduced engine emissions of sulfur oxides, carbon dioxide, particulate matter, and aromatic hydrocarbons compared to conventional diesel fuels. Due to global climate change from greenhouse gases, a decline in world crude oil production, and rising prices of petroleum products, biofuels have received considerable attention and offer promising alternate energy resources. This research being performed at Southwest Research Institute (SwRI) is an efficient continuous process to replace the costly batch method for diesel fuel production.

In the conventional transesterification of fats/vegetable oils for biodiesel production, free fatty acids and water always present a problem and result in a low conversion. Supercritical methanol (SCM) process does not have these problems. The objective of this study was to produce biodiesel continuously from waste cooking oil and animal fats in supercritical methanol (239.4^{o}C and 8.0MPa) with/without heterogeneous catalysts. The process flow diagram is shown in Figure 1.

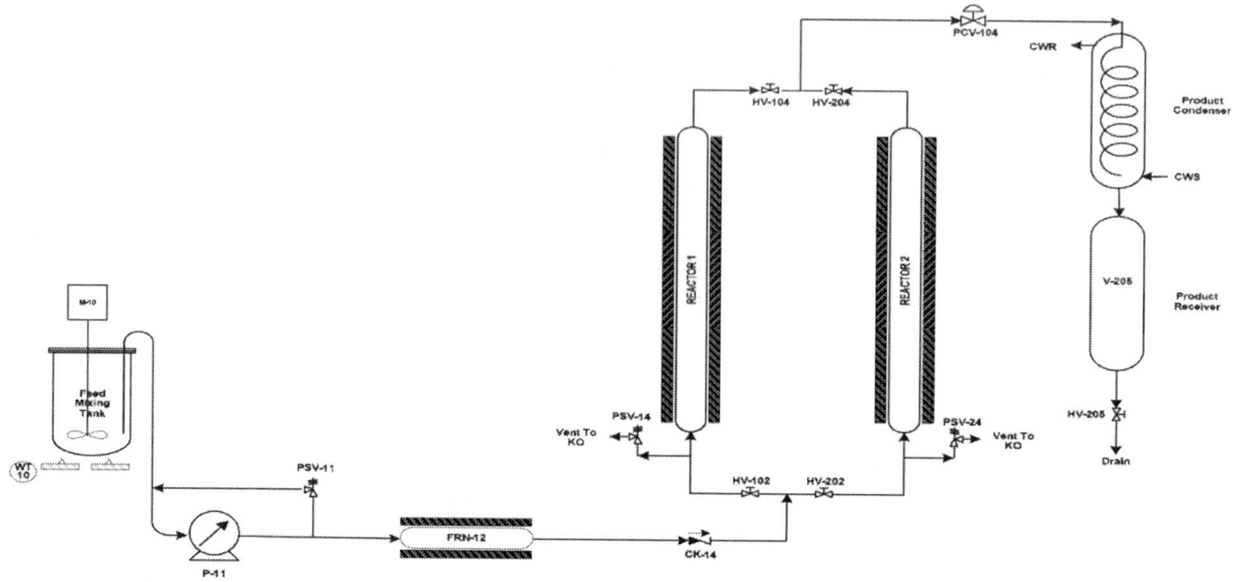

Figure 1: Process Flow Diagram for biodiesel production system.

EXPERIMENTAL SETUP

The fix-bed tubular reactor was constructed from a 3/8" ID x 4' long stainless tube with a wall thickness of 0.049" (heavy wall). The reactor has two thermocouples inserted through the side wall of the tube one-third of the way from each end to monitor the temperature of the chemicals contacting the catalyst. These thermocouples are inserted through thermowells welded to the side of each reactor tube. The pressure within the reactor system is controlled with a pressure controller and detected with a pressure transducer. A pressure gauge is also installed on-line to ensure the conditions necessary to achieve supercritical methanol are maintained.

The stainless steel reactor is installed inside an aluminum furnace enclosure insulated with a calcium silicate material. The inlet of the reactor is equipped with an on-line pressure gauge to monitor the high-pressure Eldex metering pump. Pressure safety valves are installed at the inlet of the reactors and set at 4,000 psig. A pressure exceeding the control limit will open the safety valve and release excess reactor pressure inside the hood to a safe location.

The 4L feed preparation vessel installed to mix the reagents, feedstock and methanol, was agitated with an overhead stirrer. The vessel is seated on a weigh scale so that delivery of reagents to the catalytic reactor can be recorded. At the outlet of the reactor, the pressure within the reactor is regulated with a pressure controller and a pressure transducer. These devices hold the back pressure within the reactor system to achieve supercritical methanol conditions at 239.4°C and 8.0MPa (1160 psig).

At the outlet of the pressure controller, the biofuel, fat byproducts, and excess methanol are then fed into a condenser and collected in a graduated receiver (a collection funnel and an Erlenmeyer flask). Uncondensed gases will be captured within a scrubber of activated carbon. The complete Process Flow Diagram is shown below in Figure 1. The process instrumentation is controlled by an iFix program. This software allows for data trending, automatic controls, safety alarms and fail-safe process shutdown.

EXPERIMENTAL RESULTS

GC-MS Analysis

The biodiesel analytical method was developed with a Hewlett Packard GC Model HP 6890 interfaced with an Agilent 5973 Network Mass Selective Detector. The analytical column was a DB-23 fused-silica capillary column with i.d. 0.25mm x 60m length with 0.15μm film thickness silicone-coated fused-silica capillary column. This narrow-bore capillary column is directly connected into the mass spectrometer source. The mass spectrometer is capable of scanning from 12 to 550amu many times per second, with energy of 70eV in the electron impact ionization mode. The mass spectrometer also has the capability of producing mass spectrum for decafluorotriphenylphosphine (DFTPP) as a calibration standard for daily mass tuning to ensure the ionization mass was identified correctly. The data system allows the continuous acquisition and storage on machine-readable media of all mass spectra obtained throughout the duration of the research study.

The GC-MS determines the concentrations of each compound in the product using the relative response factor (RRF) as follows:

$$Compound\ Detected\ \left(\frac{mg}{L}\right) = \frac{(RRF)compound\ detected}{(RRF)calibration\ standard}\ x\ Conc.of\ Calibration\ Standard\ \left(\frac{mg}{L}\right) \quad (2)$$

By comparing the unreacted starting reagent detected in the product to the initial concentration of feedstock in the reagent, the percent yield can be calculated. For example, the tributyrin experiments had a feedstock concentration of 158147 mg/L. The final concentration was determined by GC/MS to be 169900 mg/L. The percent conversion is obtained by taking the difference and dividing by the initial concentration. Theoretically, if there is no conversion then the final concentration of tributyrin should equal the initial concentration. Linoleic acid was assumed to be the primary component of the waste cooking oil, and the molecular weight of the oil was assumed to be 977.57 g/mol. Further testing will be conducted to determine an exact molecular weight and to identify components of triglyceride in the waste cooking oil (WCO).

The reaction conditions for transesterification of WCO under SCM condition and with/without catalyst are tabulated in Table 1, and Table 2 lists that the conversion rate for the transesterification of WCO in Table 1. Due to associated plugging problems of the highly viscous feedstock, higher concentration of pork fat was unable to perform with the current Eldex pump. Catalyst SBA-15 has been tested, but the reactor was plugged and could not be continued. Catalyst Na_2CO_3/alumina was also tested and the conversion was above 99%.

Table 1. Experimental Summary of Waste Cooking Oil (WCO) with Supercritical Methanol (SCM)

Date	Run ID	Page #	Pressure Stpt (psig)	Temp Stpt (°C)	Flow Rate (mL/min)	Methanol:Oil Ratio	Feedstock	SCM/Catalyst
09/01/10	8A	32	1500	300	2.78	40:1	WCO	SCM
09/01/10	8B	33-34	2500	300	2.40	40:1	WCO	SCM
09/02/10	9A	34-35	1500	250	2.53	40:1	WCO	SCM
09/02/10	9B	35-36	2500	250	2.20	40:1	WCO	SCM
09/02/10	9C	36	3000	250	2.40	40:1	WCO	SCM
09/02/10	9D	36-37	3000	300		40:1	WCO	SCM
09/03/10	10A	38-40	1500	325	2.17	40:1	WCO	SCM
09/03/10	10B	40-41	2000	325	2.55	40:1	WCO	SCM
09/07/10	11A	42-43	1500	325	2.40	>200:1	Pork fat	SCM
09/07/10	11B	43-44	2500	325		>200:1	Pork fat	SCM
09/14/10	13A	47	2500	250	2.87	>200:1	Pork fat	SCM
09/14/10	13B	48	3000	250	2.50	>200:1	Pork fat	SCM
09/15/10	14A	49-50	2000	300	2.93	>200:1	Pork fat	SCM
09/15/10	14B	50	2500	300	2.80	>200:1	Pork fat	SCM
09/15/10	14C	50-51	3000	300	2.60	>200:1	Pork fat	SCM
09/16/10	15A	52	2000	200	2.47	>200:1	Pork fat	SCM
09/16/10	15C	53-54	3000	200	1.95	>200:1	Pork fat	SCM
09/24/10	16A	56	1500	200	2.50	40:1	WCO	SCM
09/24/10	16B	58	2000	200	2.00	40:1	WCO	SCM
09/30/10	20A	64	1500	250	3.08	40:1	WCO	SCM
10/13/10	25A	81-82	2500	250	4.70	40:1	WCO	SCM
10/13/10	25B	82	2500	250	5.57	40:1	WCO	SCM
10/13/10	25C	82-83	2500	250	7.00	40:1	WCO	SCM
10/21/10	28A	91	2000	250	2.83	40:1	WCO	SCM
10/21/10	28B	92	1700	250	2.86	40:1	WCO	SCM

10/21/10	28C	92	2200	250	3.00	40:1	WCO	SCM

*40:1 molar ratio of MeOH:WCO is equivalent to (977.57g WCO / 2.68 L total volume) = 365,000 mg/L.

Table 2: Conversion Rate for Biodiesel Production from Waste Cooking Oil

Sample ID	Run ID	SCM/Catalyst	mg/mL	mg/L	% Conversion
8A-033	8A	SCM	33.0	33000	**88.2**
8B-035	8B	SCM	0.3	340	**99.9**
9A-037	9A	SCM	7.0	7000	**97.5**
9B-039	9B	SCM	0.9	930	**99.7**
9C-041	9C	SCM	0.6	560	**99.8**
9D-043	9D	SCM	0.6	560	**99.8**
10A-045	10A	SCM	0.5	540	**99.8**
10B-046	10B	SCM	0.6	580	**99.8**
11A-049	11A	SCM	8.2	8200	**97.1**
11B-050	11B	SCM	0.6	590	**99.8**
16A-062	16A	SCM	1.5	1500	**99.5**
16B-063	16B	SCM	15.0	15000	**94.6**
20A-064	20A	SCM	1.4	1400	**99.5**
21A-066	21A	Zeo Y	3.7	3700	**98.7**
21B-067	21B	Zeo Y	16.0	16000	**94.3**
22A-068	22A	Zeo X	2.3	2300	**99.2**
22C-071	22C	Zeo X	2.9	2900	**99.0**
24A-072	24A	Zeo Y	2.5	2500	**99.1**
24B-073	24B	Zeo Y	9.7	9700	**96.5**
24C-074	24C	Zeo Y	77.0	77000	**72.5**
24D-075	24D	Zeo Y	106.0	106000	**62.1**
25A-076	25A	SCM	17.0	17000	**93.9**
25B-077	25B	SCM	26.0	26000	**90.7**
25C-078	25C	SCM	38.0	38000	**86.4**
26A-079A	26A	Zeo Y	32.0	32000	**88.6**
26B-079B	26B	Zeo Y	14.0	14000	**95.0**
26C-080	26C	Zeo Y	16.0	16000	**94.3**
27A-081	27A	Zeo+H_2SO4	157.0	157000	**43.9**
27B-082	27B	Zeo+H_2SO4	8.2	8200	**97.1**
28A-083	28A	SCM	3.6	3600	**98.7**
28B-084	28B	SCM	2.2	2200	**99.2**

28C-085	28C	SCM	18.0	18000	**93.6**

Note: Waste cooking oil (WCO) contained 280,000 mg/L of triglyceride (76.7% triglyceride).

SUMMARY

Transesterification of waste cooking oil (WCO) and animal fat with methanol under supercritical condition gave high yield of fatty acid methyl ester (FAME) (>90%) under the following reaction conditions: reaction temperatures = 300°C and 250°C; reaction pressures = 2000 psi, 2500 psi, and 3000 psi; residence time = 18 minutes; molar ratio of methanol to oil = 40:1; and flow rate = 2.5 mL/min. At 250°C and 2500 psi, when the WCO flow rate was increased to 4.5mL/min and 6.5mL/min, and FAME yields were >90% for the flow rates. The findings demonstrated that, by a supercritical methanol and solid catalytic approach, waste cooking oil and animal fat could be readily used for biodiesel fuel production in a simple, continuous, economically effective, and automatic process.

(110g) Chemical Components That Influence Evaporator Distillate Quality

Shawn D. Thornton, R & D, HPD, LLC, Plainfield, IL

The evaporation process is a very effective separation method for high-strength brines. This presentation is focused on the typical components that can appear in evaporator distillate and the key factors that can influence distillate quality. Testing experience regarding the quality of evaporator distillate is discussed.

In this presentation the term "evaporator distillate" refers to evaporator condensate. It is important to consider where the evaporator distillate will be used when evaluating its quality. Some uses of evaporator distillate are discharge to surface water, process reuse, and boiler water make-up.

There are several compounds and groups of compounds of interest in evaporator distillate quality. There is a marked tendency for both ammonia and bicarbonate to report to evaporator distillate. This tendency has a strong influence on both the conductivity and pH of the evaporator distillate. The influence of evaporator pH on the reporting of both ammonium bicarbonate and free ammonia to the evaporator distillate is discussed. A discussion of total organic carbon (TOC) and several organic components that can report to evaporator distillate is provided. Fingerprinting of semivolatile organic components in an evaporator feed and the corresponding evaporator distillate sample using gas chromatography analysis is illustrated. Results are provided showing that trace amounts of boron can report to evaporator distillate.

A method to improve evaporator distillate quality is vapor washing. The concentrations of non-volatile inorganic components in evaporator distillate are typically very low, so the concept of decontamination factor is used. Vapor washing can improve the decontamination factors for non-volatile inorganic components. Foaming is an upset condition that can be managed with the use of a vapor washer.

AUTHOR INDEX